CW01305735

small stuff

small stuff

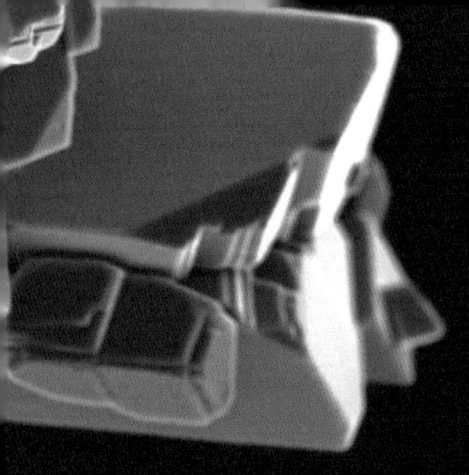

Colorized Scanning Electron Microscopy

Mark Trosper McClendon

High Resolution Prints Available Here:
fineartamerica.com/profiles/mark-mcclendon.html

Printed by CreateSpace
Copyright © 2017 Mark Trosper McClendon

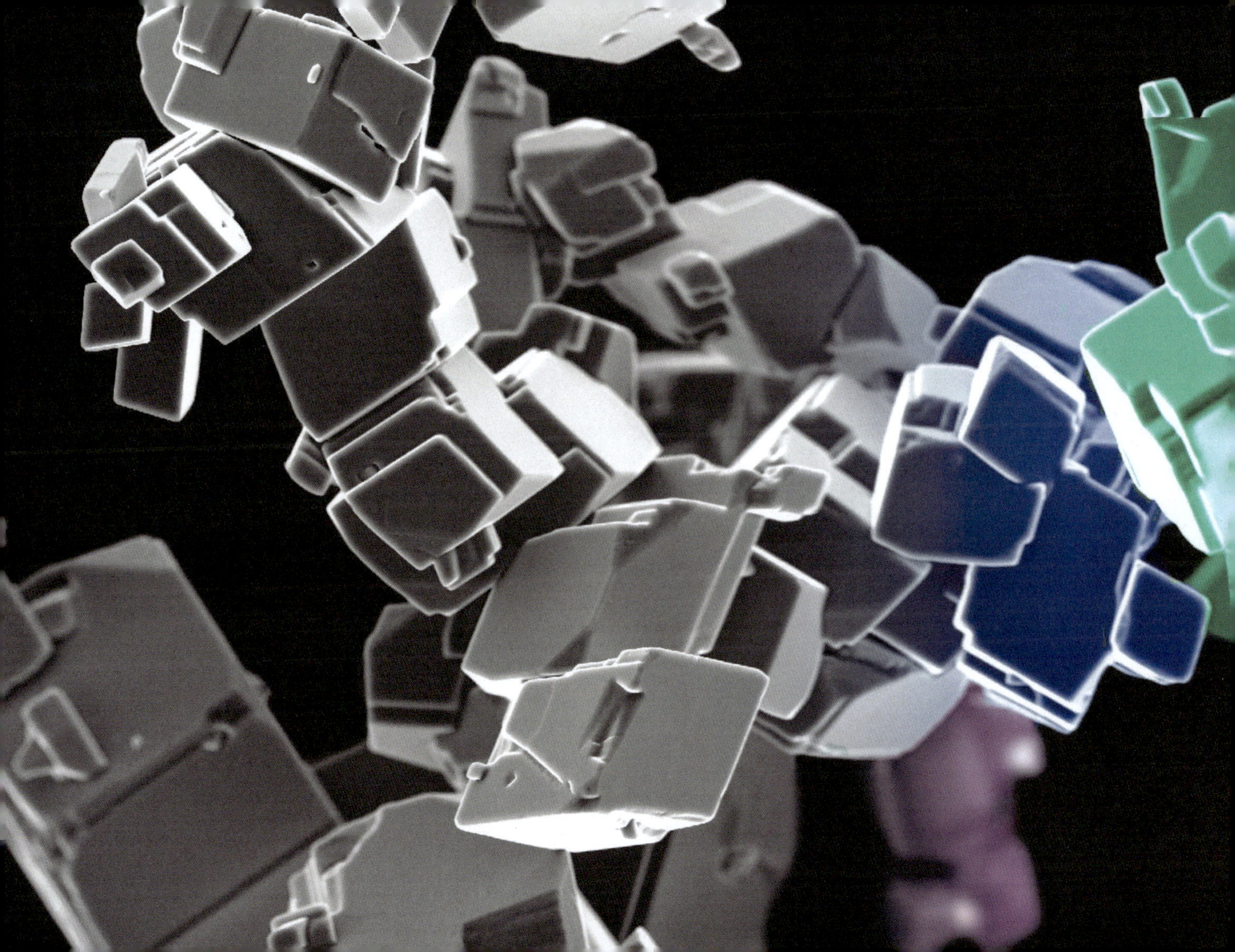

Acknowledgements

I would like to thank the following people for their help in making this book. Professor Samuel I. Stupp for his guidance through my PhD studies. The Simpson Querrey Institute for access to world class research facilities. Northwestern's NUANCE staff and Professor Vinayak P. Dravid for their training and oversight during my experiments. Michael Kennedy and Science in Society for their continued support of academic outreach through scientific images.

A special thanks to all the scientists that provided me with samples. I cannot take credit for all of the science represented in this book. I would like to thank the following scientists who were primarily responsible for these projects. I am sincerely grateful for the opportunity to work with you; Dr. Christopher Volker Synatschke [page 71], Dr. Stacey Chin [page 71], Dr. Ben Cosgrove [page 101], Dr. Eduard Sleep [page 101, 110, 111], Dr. Zaida Alvarez Pinto [page 102-105], Dr. Timmy Fyrner [page 108], Dr. Ronit Freeman [page 114,115], Dr. Sungsoo (Seth) Lee [page 108, 116], Dr. Wei Ji [page 106]

And most importantly thank you to my mom, Kim McClendon, for everything.

This work made use of the (EPIC, Keck-II, and/or SPID) facility(ies) of the NUANCE Center at Northwestern University, which has received support from the Soft and Hybrid Nanotechnology Experimental (SHyNE) Resource (NSF NNCI-1542205); the MRSEC program (NSF DMR-1121262) at the Materials Research Center; the International Institute for Nanotechnology (IIN); the Keck Foundation; and the State of Illinois, through the IIN.

Forward

This book is intended for anyone that wants to see beyond the capabilities of the naked eye. The scientific advances of the 20th century have given us the opportunity to view the microscopic world with astounding clarity. Simple things from everyday life are incredibly different, complex, and beautiful from a microscopic vantage point. The purpose of this book is to provide a small glimpse into the unseen beauty surrounding us at every moment.

The majority of these images were taken during my PhD studies at Northwestern. SEM was a primary tool for my research projects, and I used any extra microscope time to observe what I considered "fun samples". After about 5 years I amassed quite a collection and finally assembled them into this book. If there is one thing I've learned from microscopy, it's that anything is amazing if you look close enough.

A Note to the Reader

All images in this book were taken with a Scanning Electron Microscope (SEM). These microscopes work by bouncing electrons off samples rather than using light. Electrons don't have color so all SEM images will always be black and white like the image below on the left. The color is digitally added to distinguish different materials and for artistic appeal.

Fresh off the Microscope

After Colorization

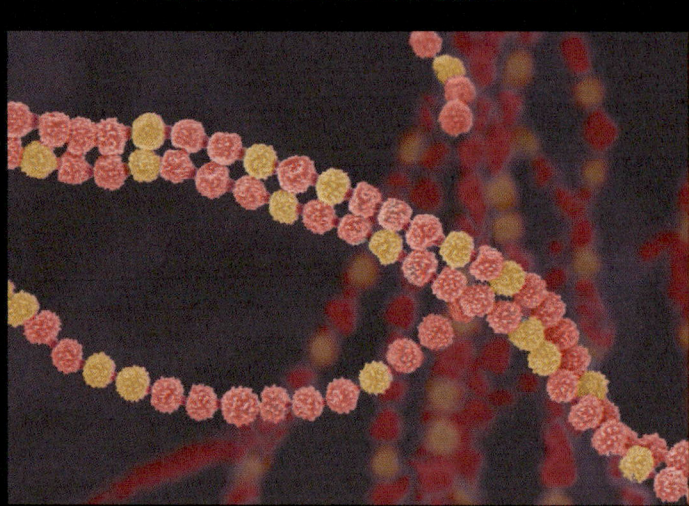

Table of Contents

Pollen 2
Insects and Arachnids 18
Creatures 46
Common Stuff 58
Bacteria, Fungus, and Plants 68
Minerals 84
Cells and Science 104
About the Author 120

Pollen
(a.k.a. Plant Sperm) is a fine powdery substance composed of very small grains with features too small to be viewed by the human eye. The variety of shapes and sizes is impressive when observed on the microscopic scale.

5μm

Cutleaf Rosinweed
(Silphium pinnatifidum)
North America

Lobster Claw Ruellia
(Ruellia colorata)
Brazil

Tropical Cactus
(Rhipsalis pachyptera)
Brazil

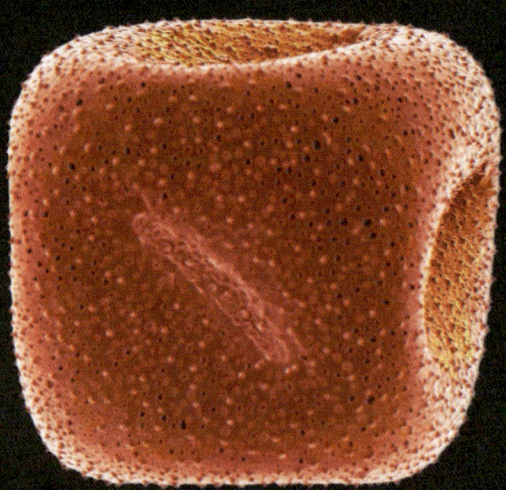

Gold Forsythia Shrub
(Forsythia x intermedia)
Asia

Lion's Ear
(Leonotis leonurus)
South Africa

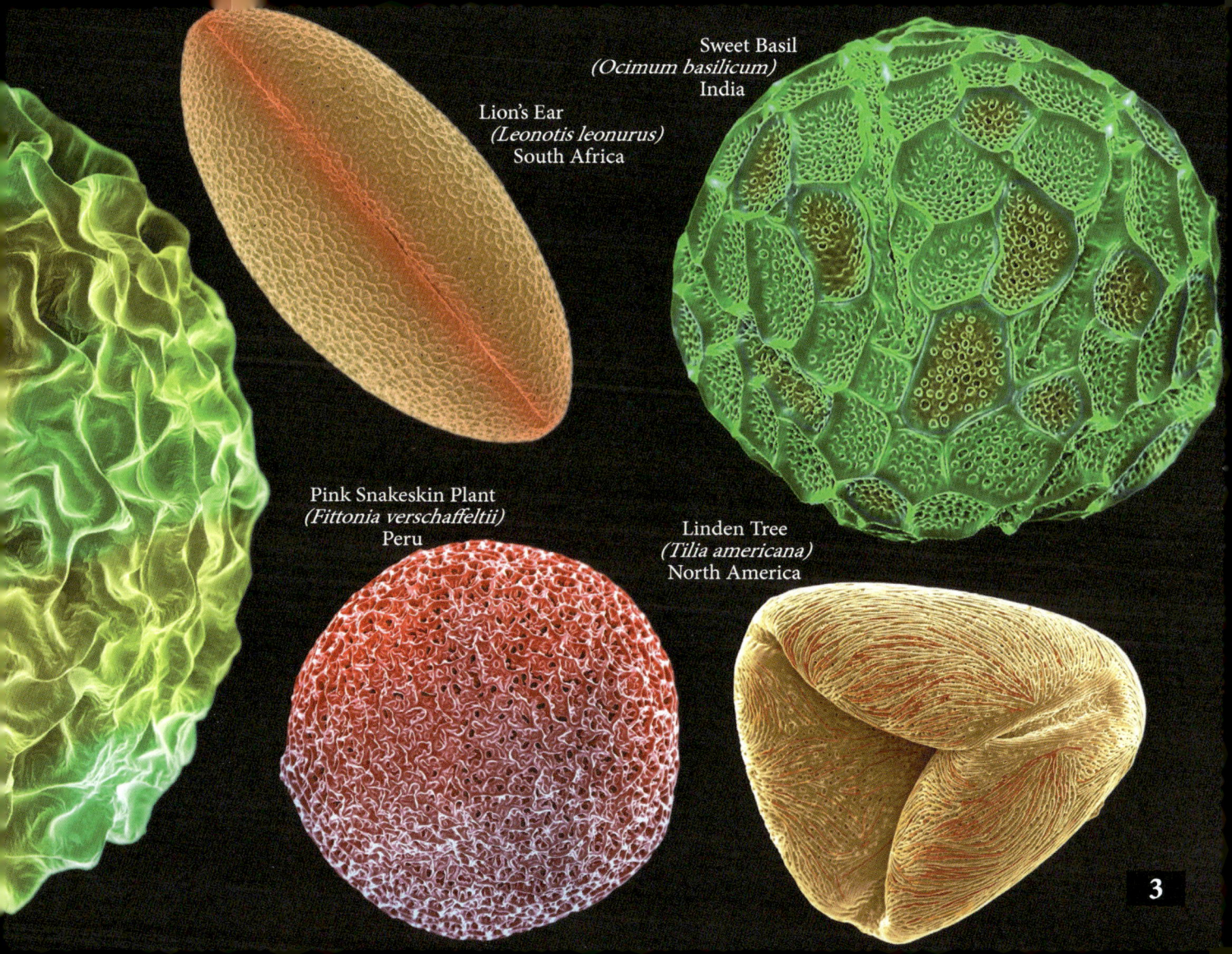

Sweet Basil
(Ocimum basilicum)
India

Pink Snakeskin Plant
(Fittonia verschaffeltii)
Peru

Linden Tree
(Tilia americana)
North America

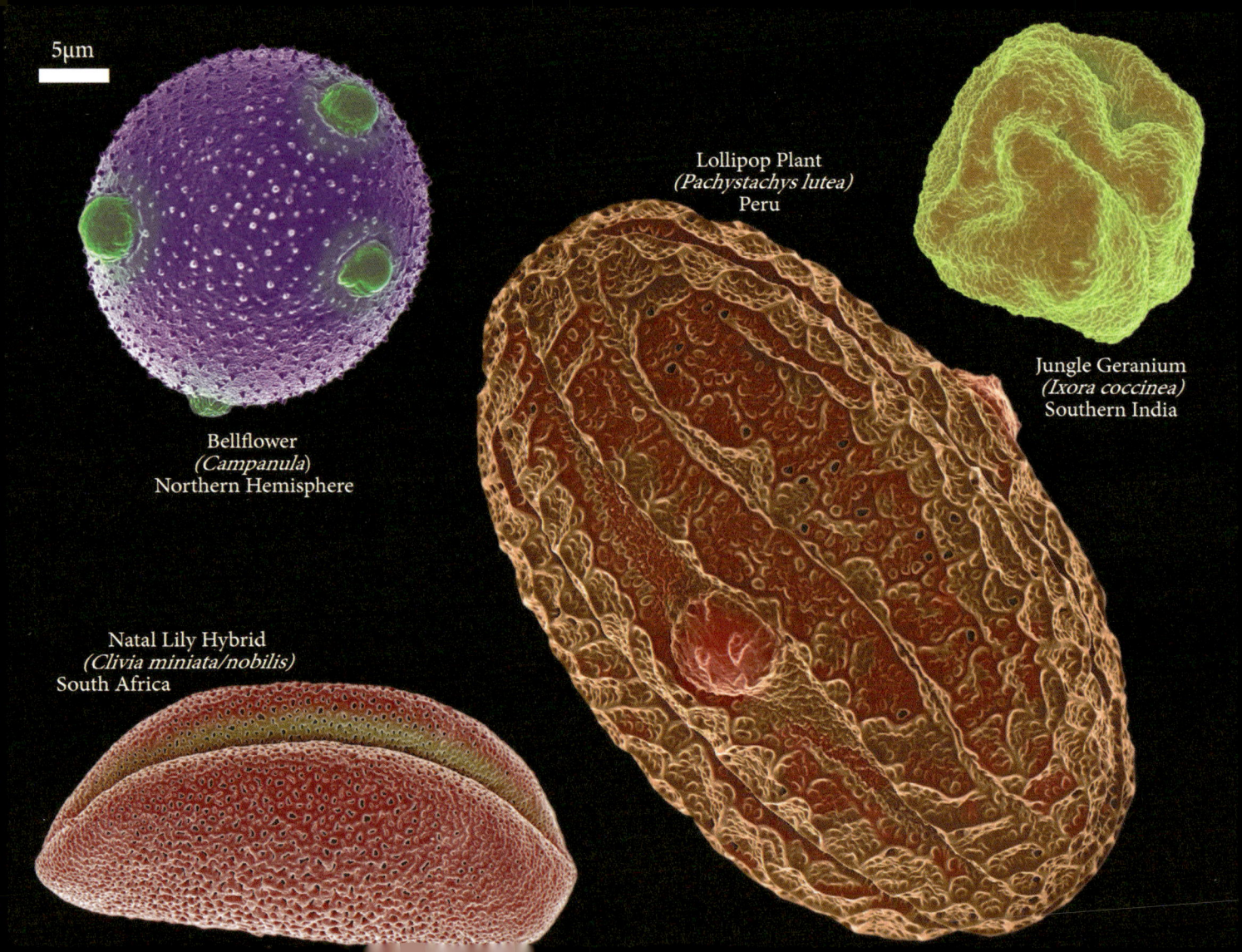

5μm

Bellflower
(*Campanula*)
Northern Hemisphere

Lollipop Plant
(*Pachystachys lutea*)
Peru

Jungle Geranium
(*Ixora coccinea*)
Southern India

Natal Lily Hybrid
(*Clivia miniata/nobilis*)
South Africa

Brazilian Plume
(Justicia carnea)
South Americal

Lobster Claw Ruellia
(Ruellia colorata)
Brazil

Cigar Plant
(Cuphea micropetala)
Mexico

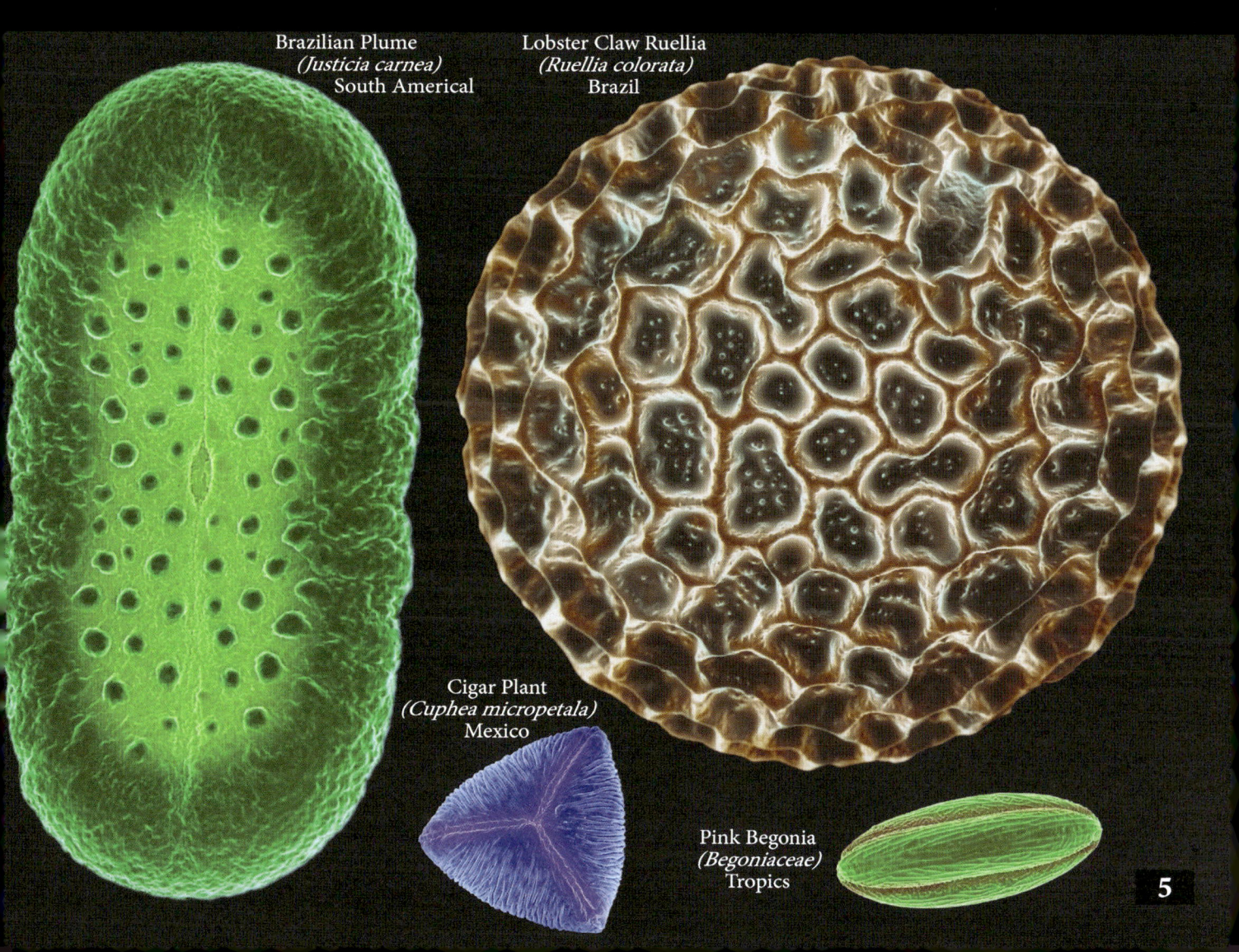

Pink Begonia
(Begoniaceae)
Tropics

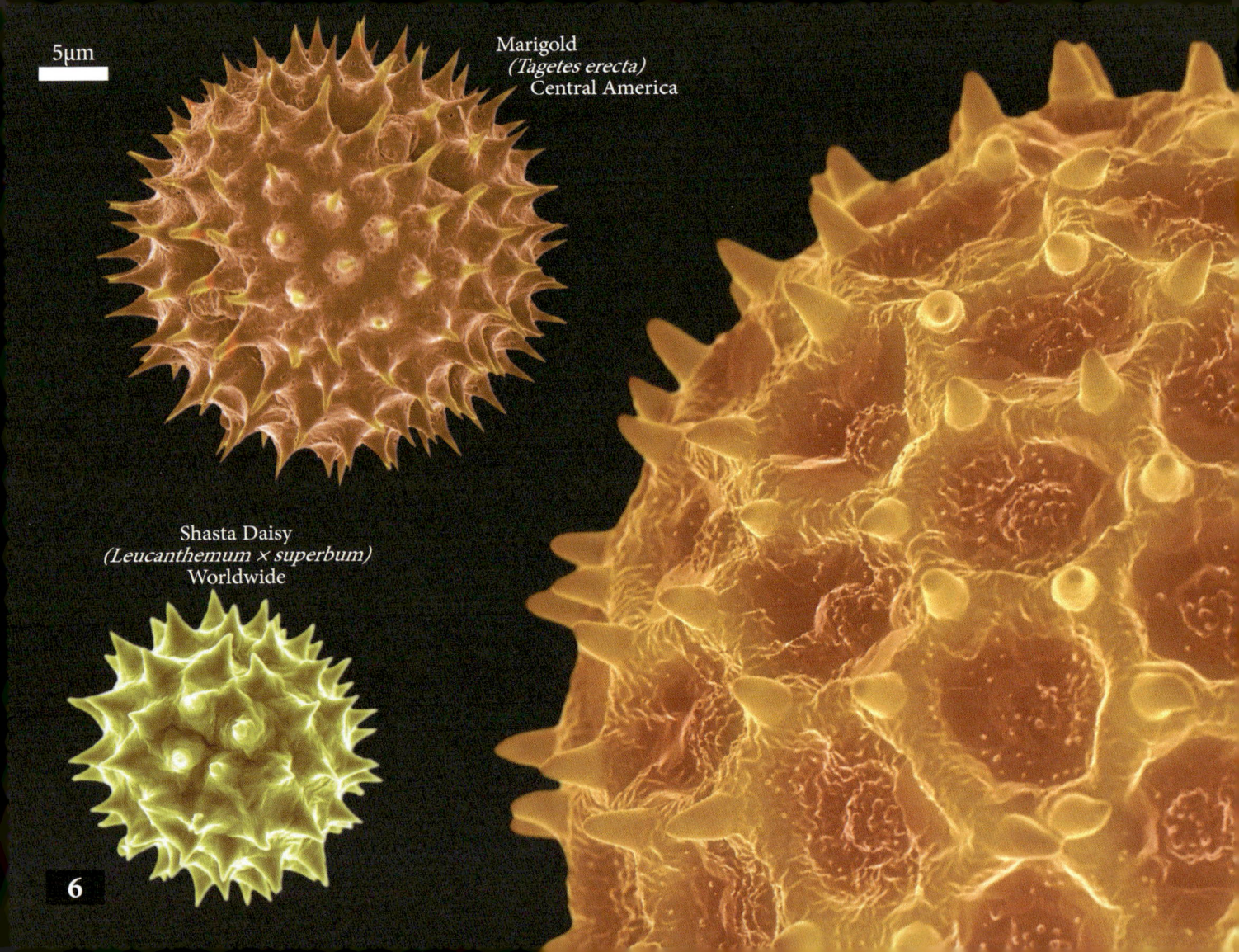

5μm

Marigold
(Tagetes erecta)
Central America

Shasta Daisy
(Leucanthemum × superbum)
Worldwide

6

Purple Morning Glory
(Ipomoea nil)
Tropics

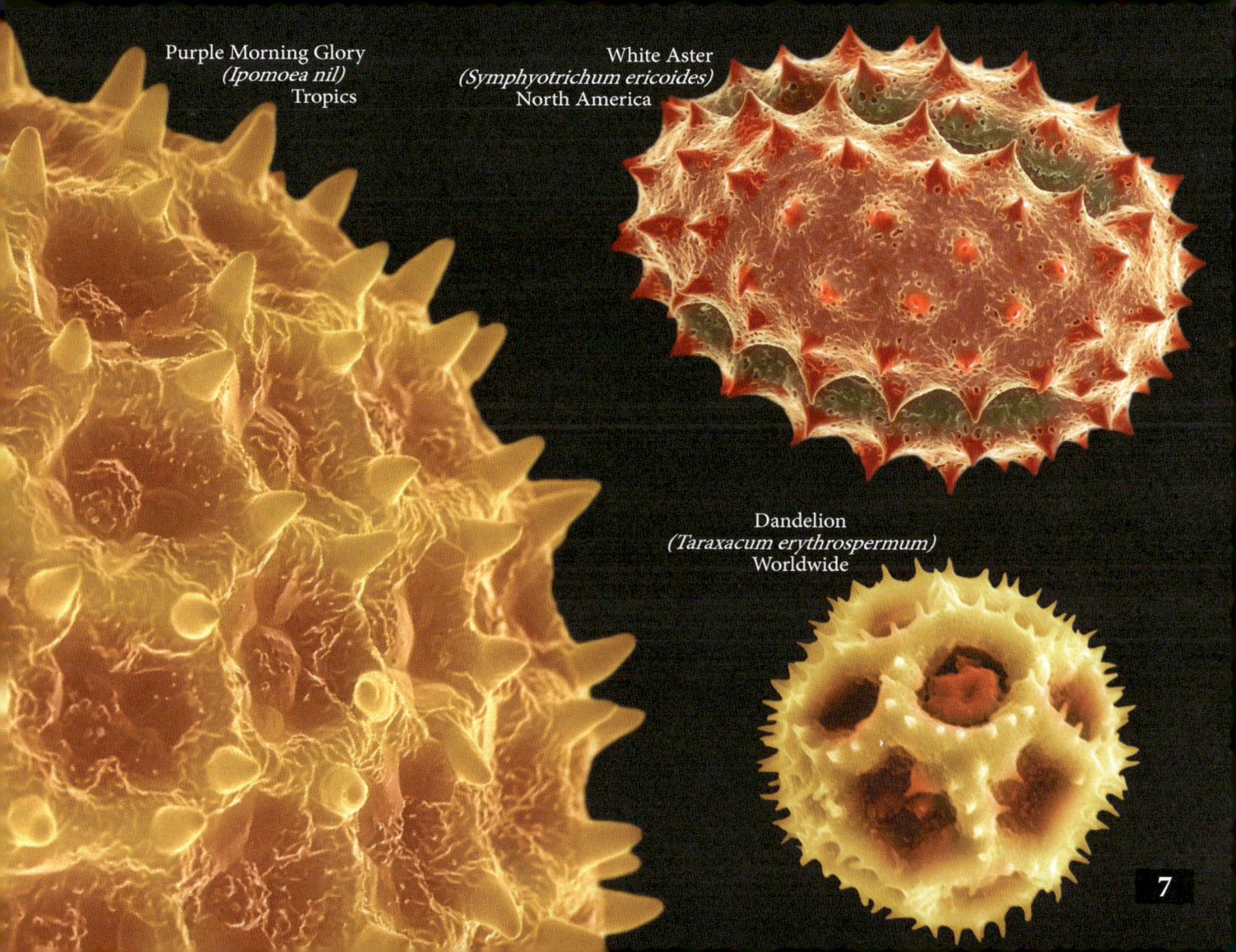

White Aster
(Symphyotrichum ericoides)
North America

Dandelion
(Taraxacum erythrospermum)
Worldwide

7

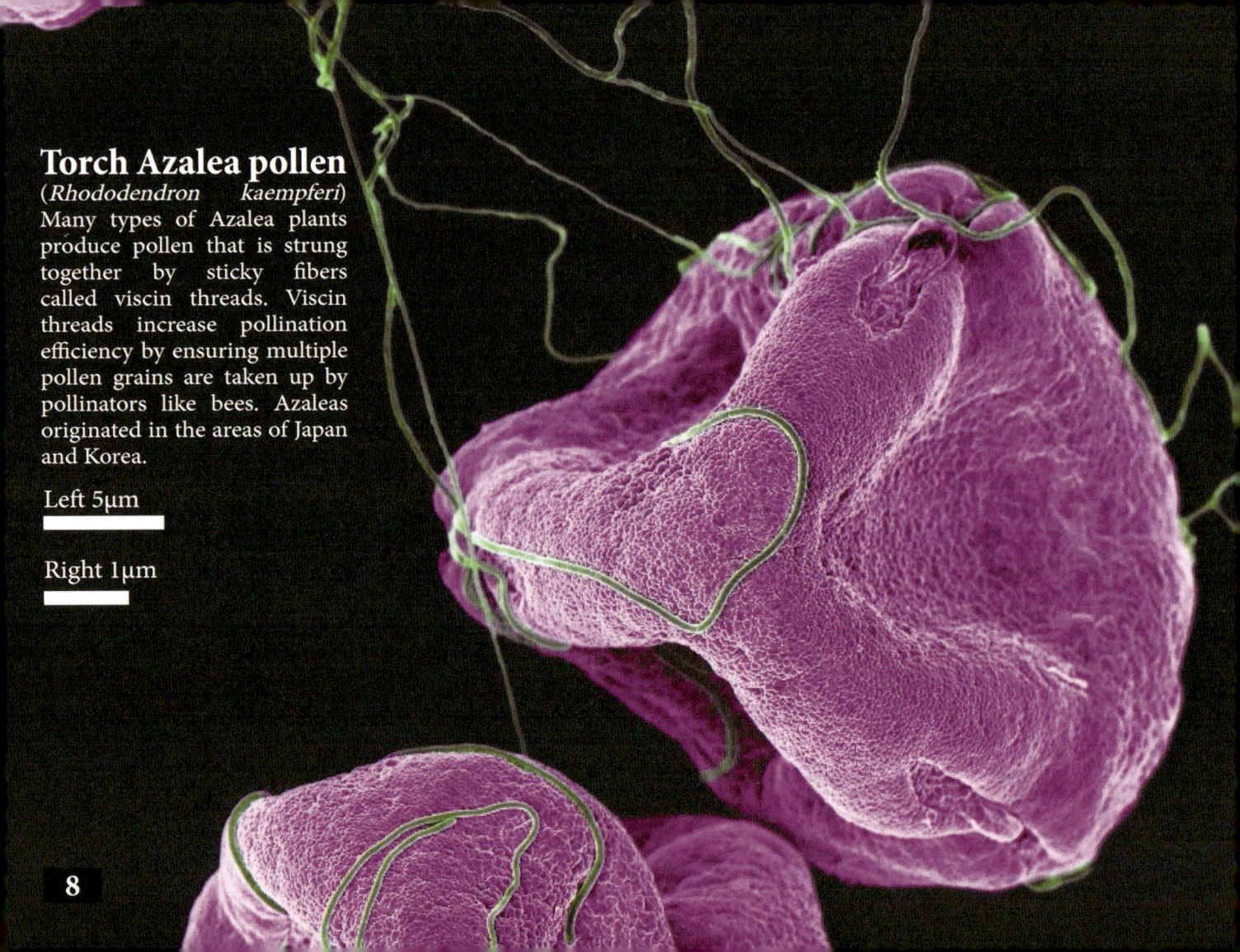

Torch Azalea pollen
(*Rhododendron kaempferi*)
Many types of Azalea plants produce pollen that is strung together by sticky fibers called viscin threads. Viscin threads increase pollination efficiency by ensuring multiple pollen grains are taken up by pollinators like bees. Azaleas originated in the areas of Japan and Korea.

Left 5μm

Right 1μm

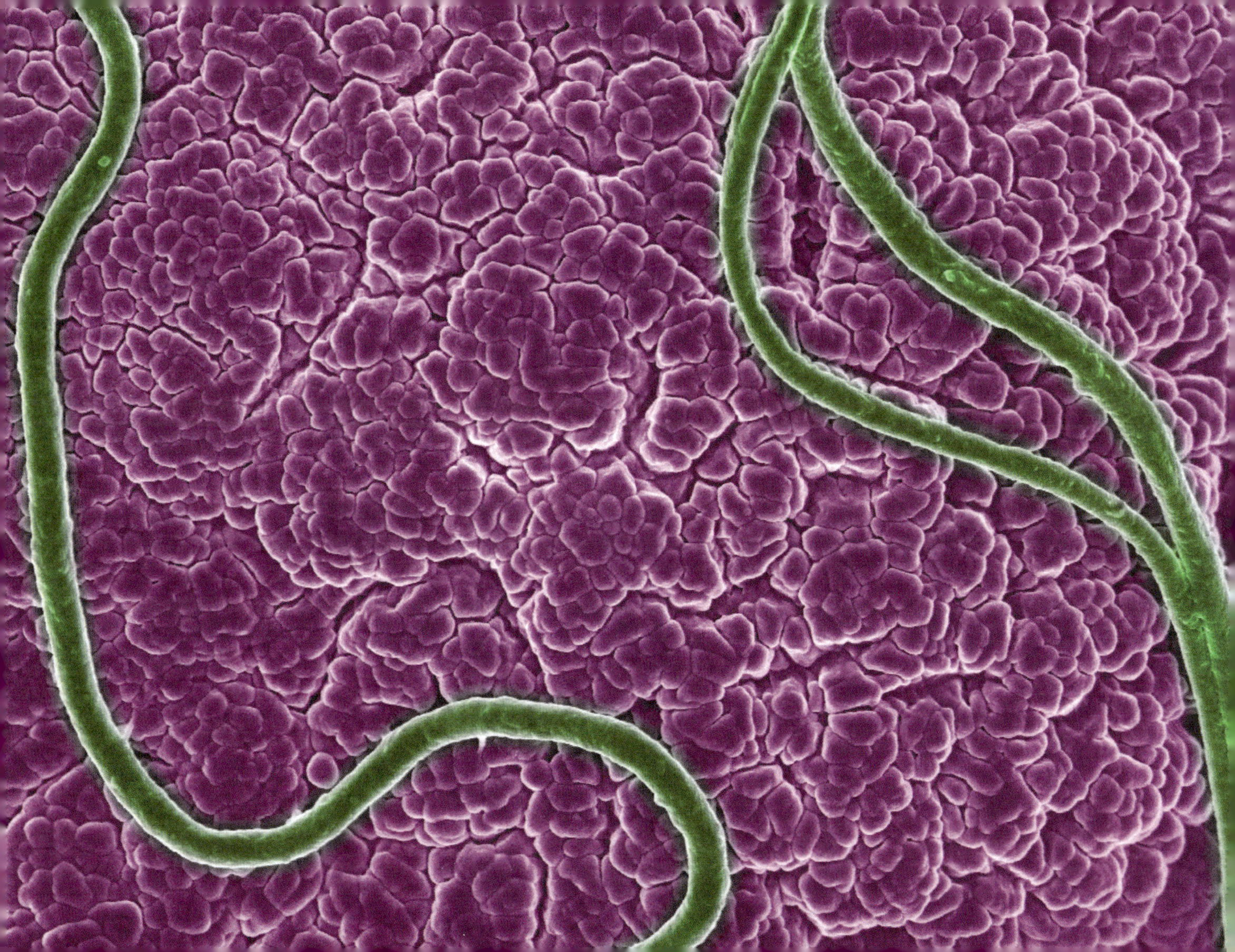

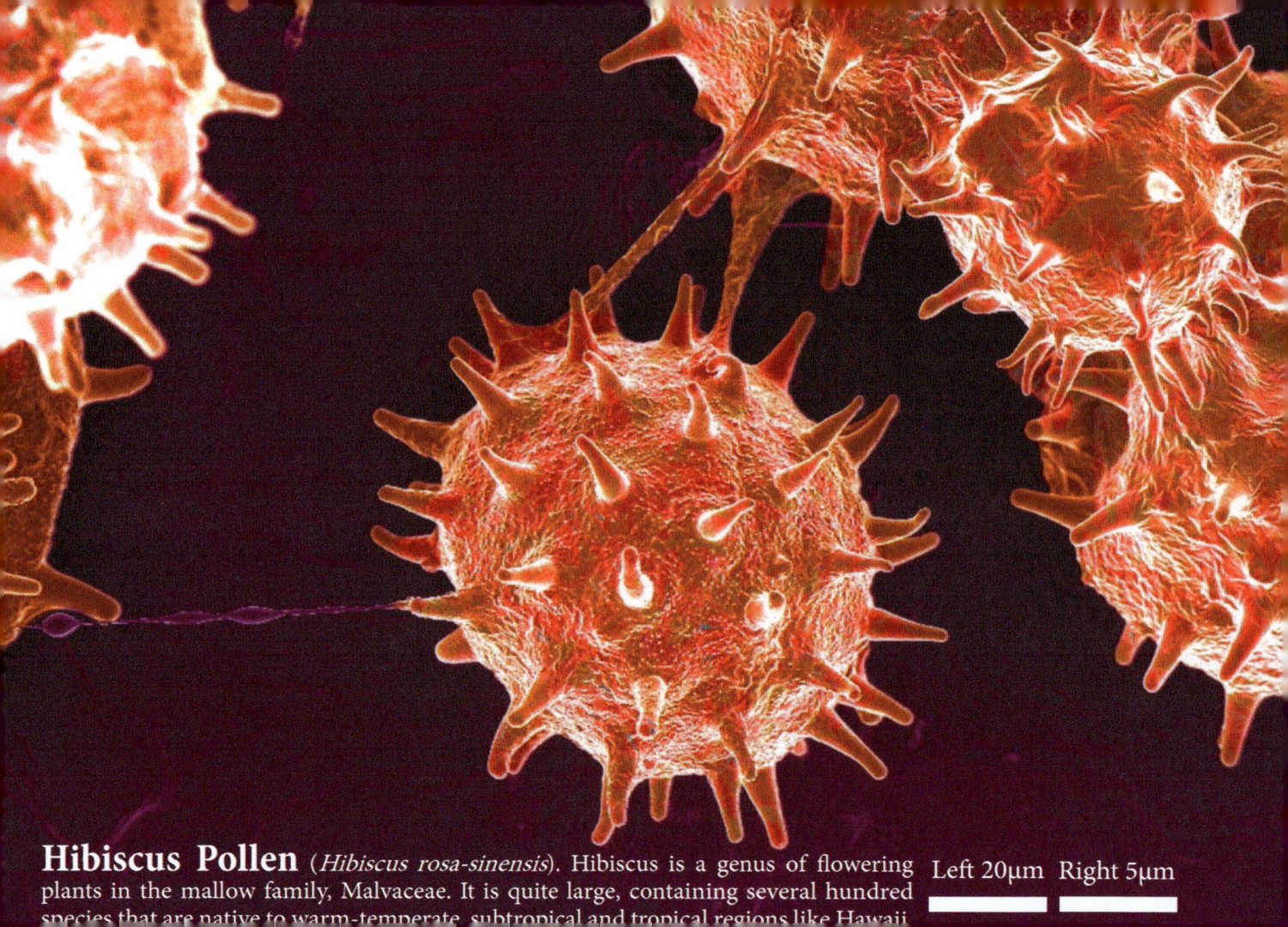

Hibiscus Pollen (*Hibiscus rosa-sinensis*). Hibiscus is a genus of flowering plants in the mallow family, Malvaceae. It is quite large, containing several hundred species that are native to warm-temperate, subtropical and tropical regions like Hawaii

Left 20μm Right 5μm

11

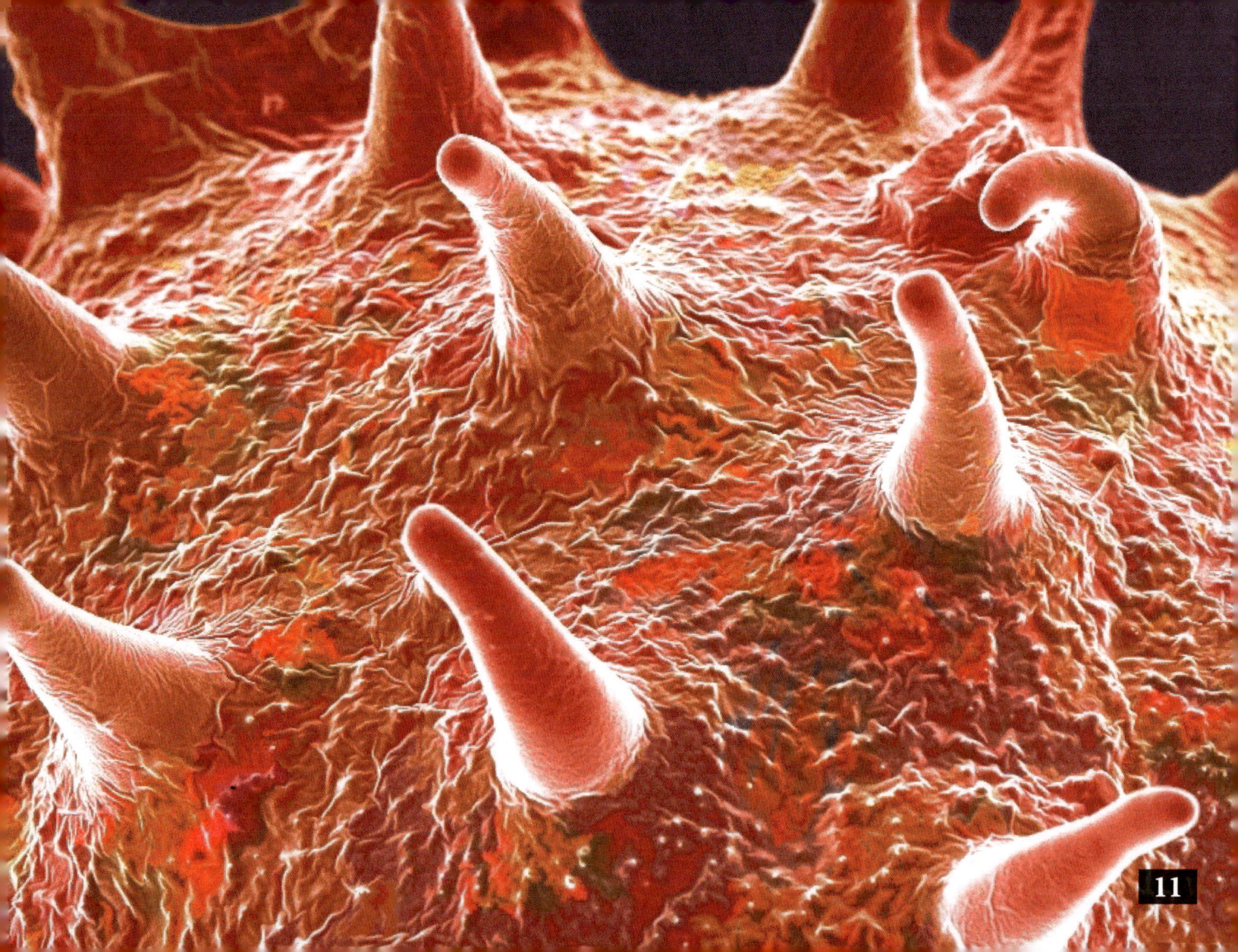

Dandelion Pollen

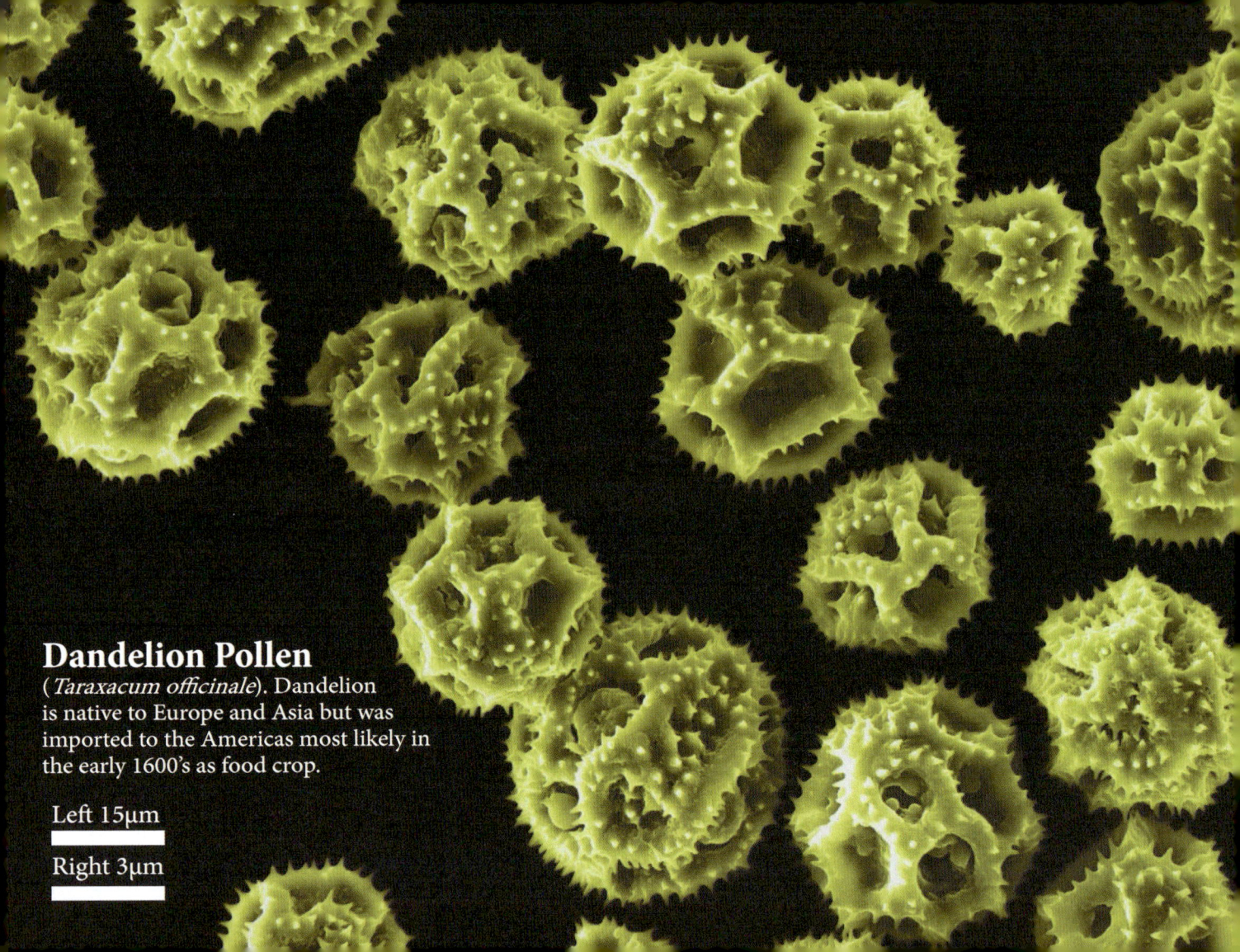

(*Taraxacum officinale*). Dandelion is native to Europe and Asia but was imported to the Americas most likely in the early 1600's as food crop.

Left 15μm

Right 3μm

13

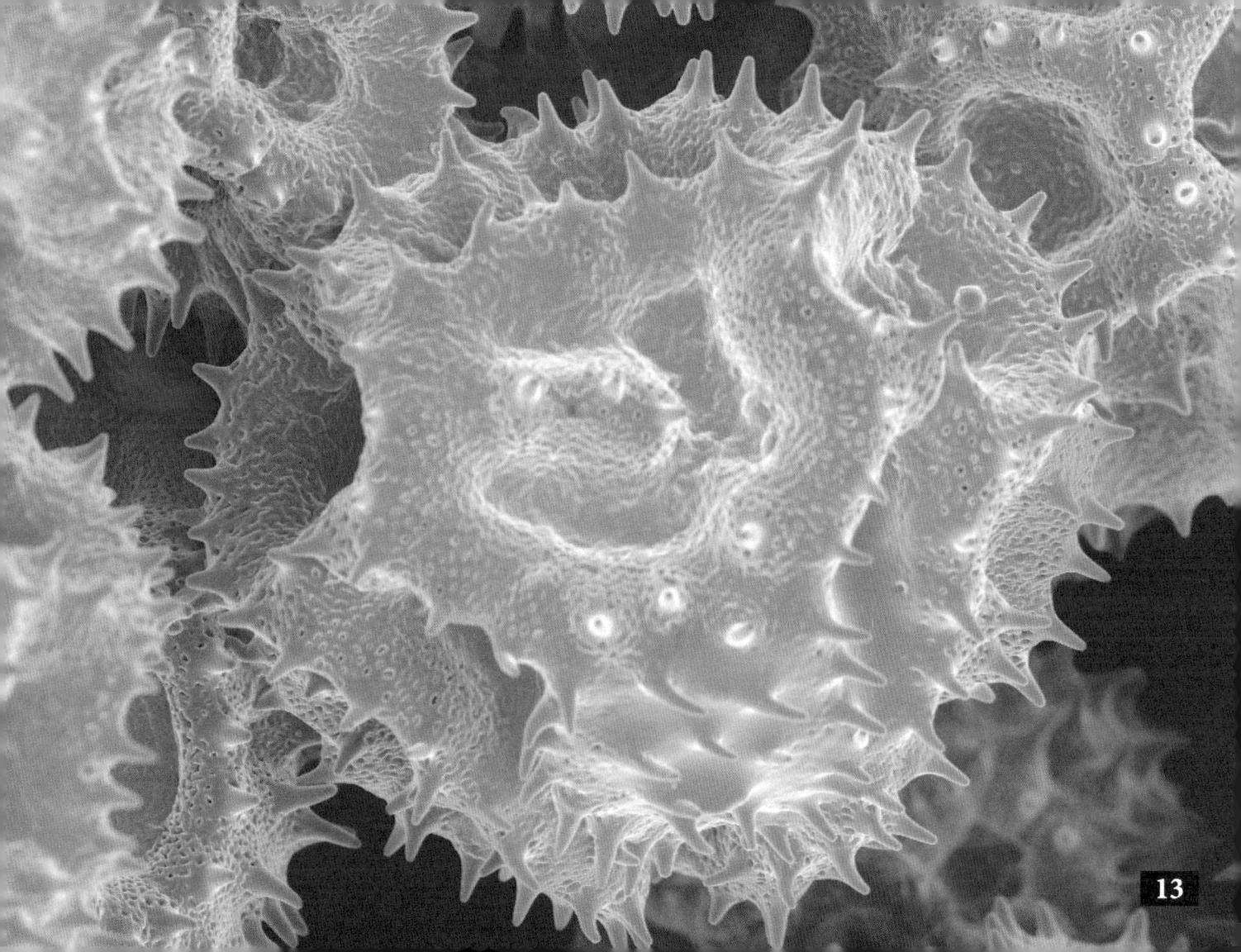

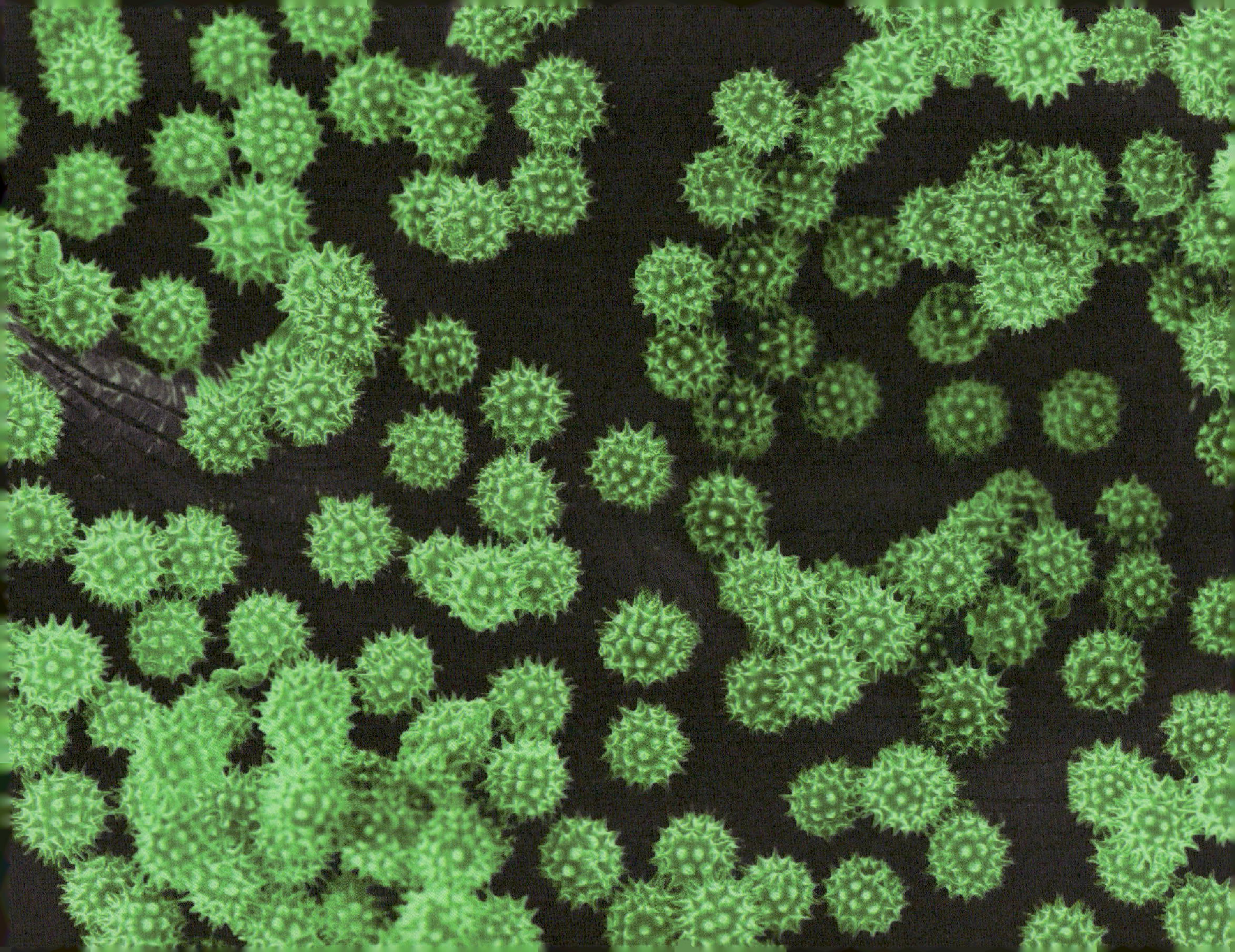

Shasta Daisy Pollen (*Leucanthemum × superbum*)

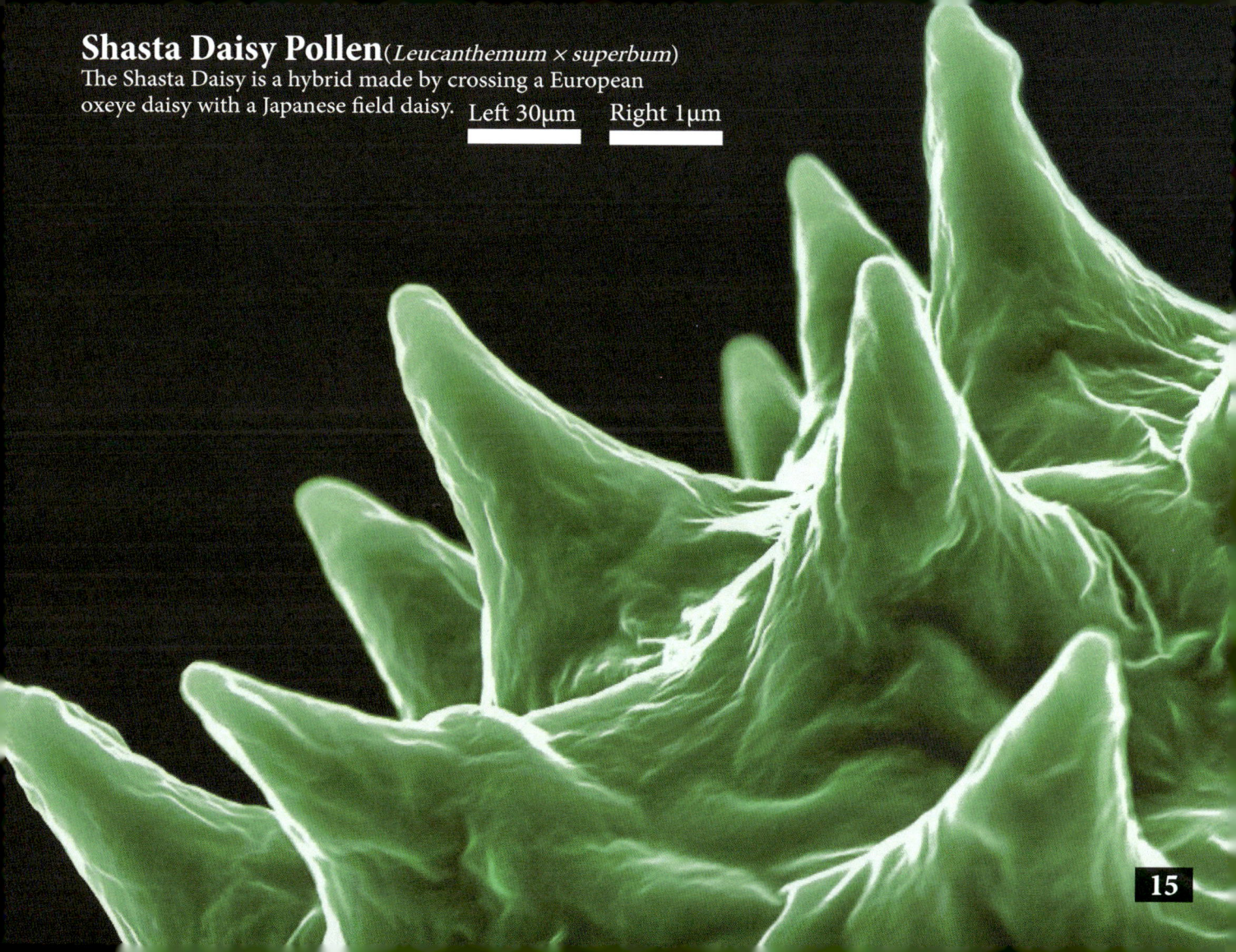

The Shasta Daisy is a hybrid made by crossing a European oxeye daisy with a Japanese field daisy.

Left 30μm   Right 1μm

15

Linden Tree Pollen
(Tilia americana) 5µm

16

Cherry Blossom Pollen
(*Prunus serrulata*) 15μm

INSECTS & ARACHNIDS

Moth scales serve many purposes such as insulation and colorful patterns. The scale's ability to easily detach from the moth makes it possible to escape some predators and spider webs.

20μm

Drain Flies (*Psychodidae*) can be found in damp environments like bathrooms and kitchens. While their fury appearance resembles a moth they are technically flies. There are over 4,700 known species worldwide.

100μm

15μm

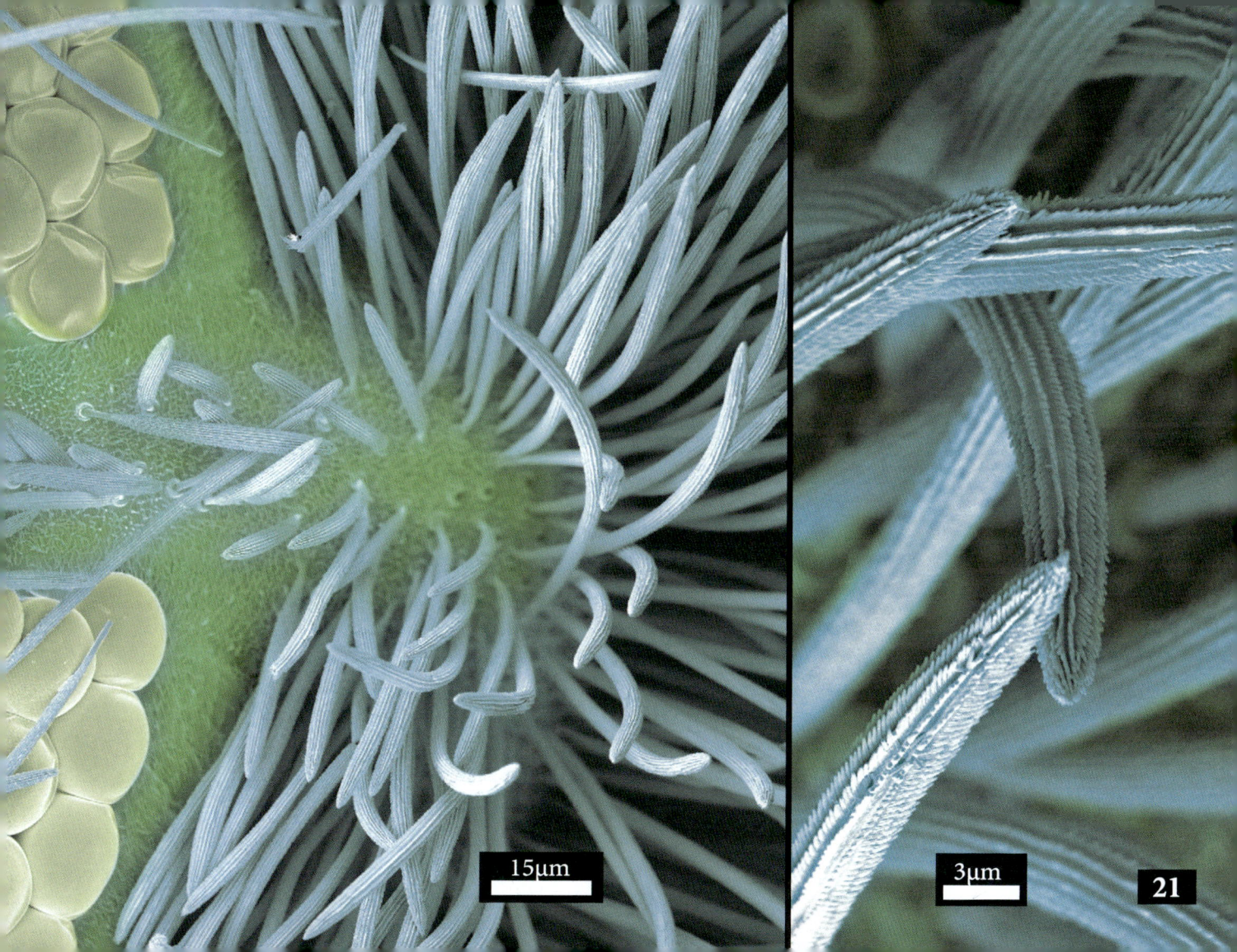

3μm

21

Fruit Flies

(*Drosophilidae*), like many flying insects, have compound eyes along with three simple eyes on the top of the head called ocelli. These simple eyes do not view images, but instead they are used to detect small changes in light allowing for flight corrections.

Right 40μm

Right 15μm

23

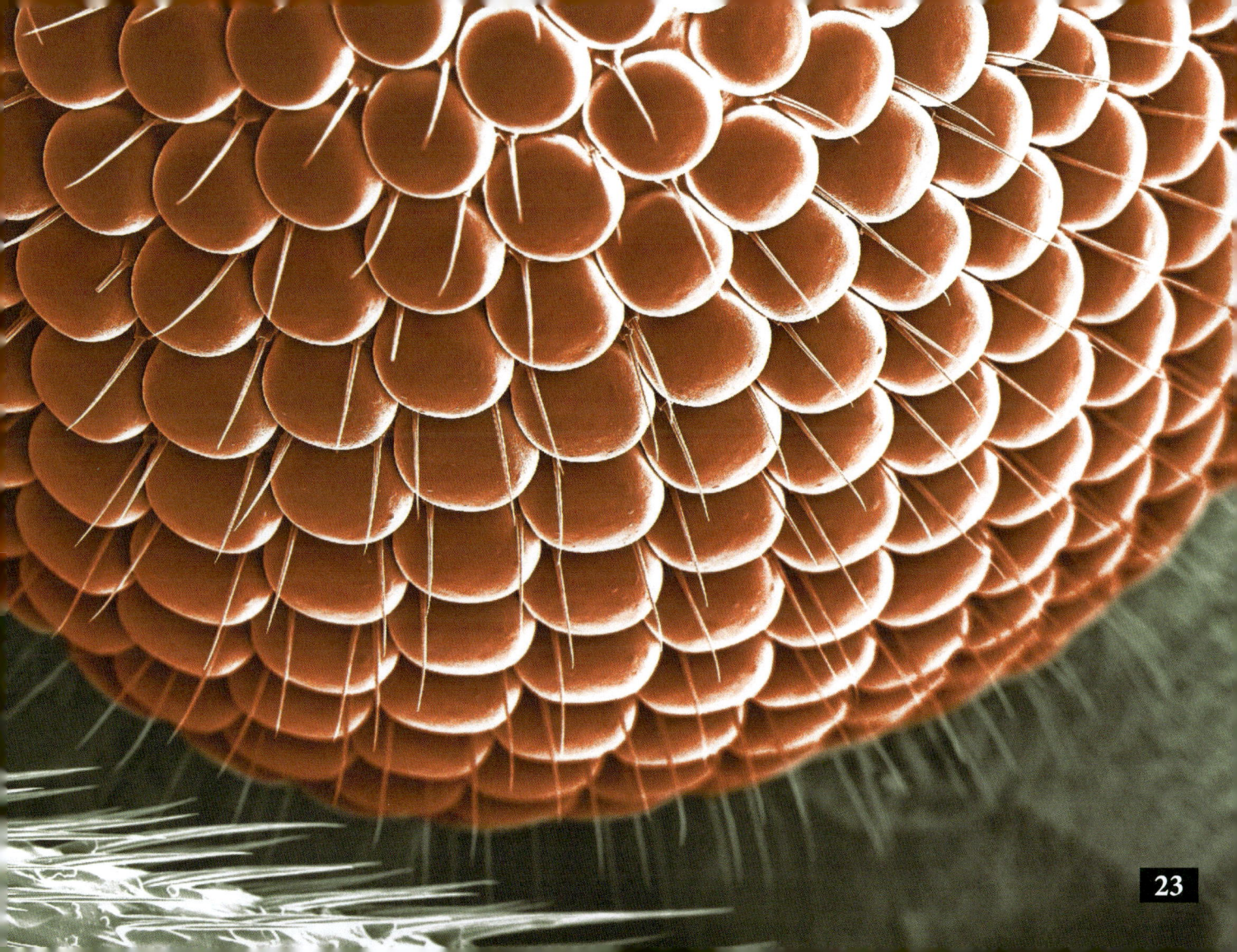

The **Red Spider Mite** (*Tetranychus urticae*) is considered a pest by most gardeners of indoor and outdoor plants. They feed on almost all food crops and many decorative plants. 100µm

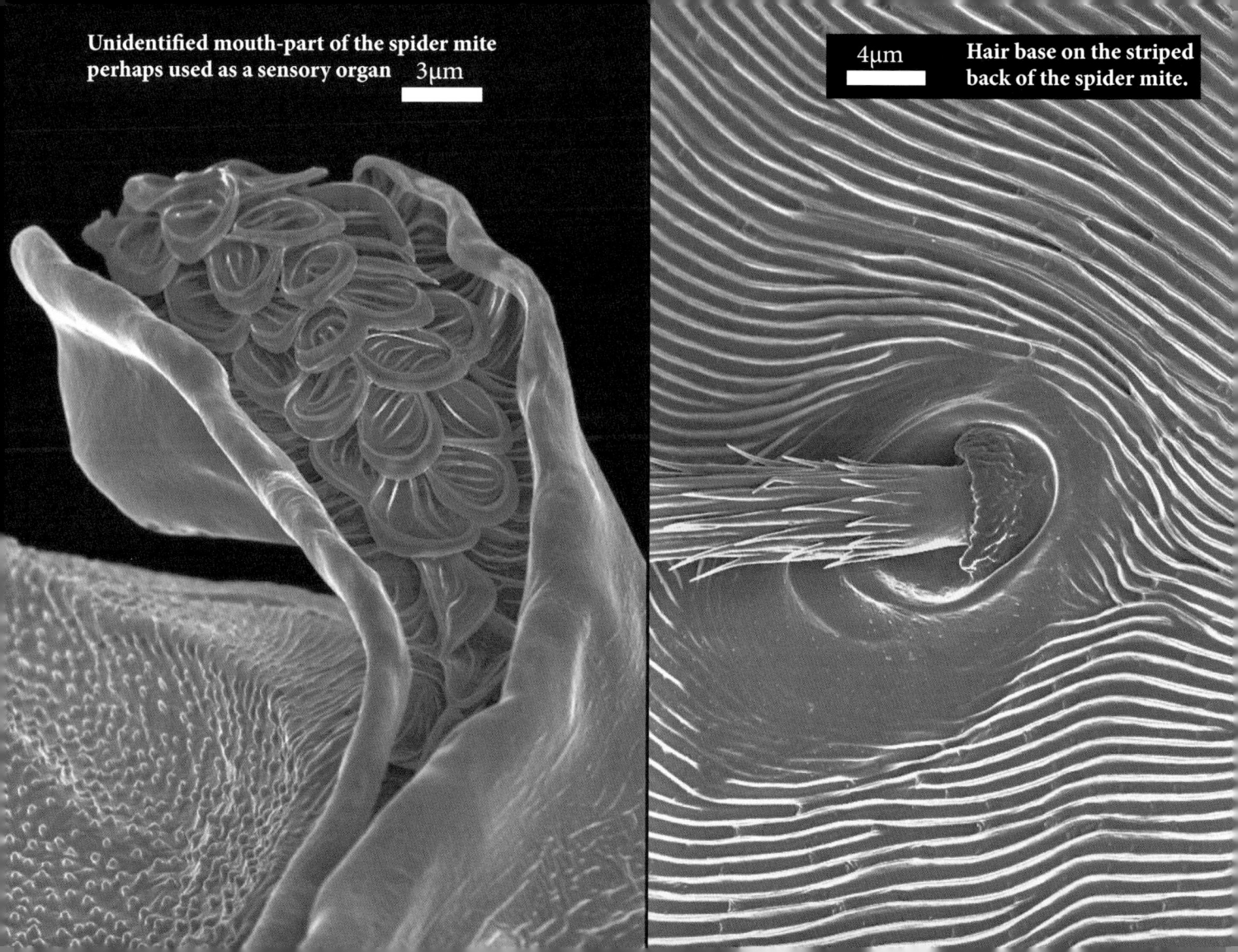

Unidentified mouth-part of the spider mite perhaps used as a sensory organ   3μm

4μm   **Hair base on the striped back of the spider mite.**

Mosquitos (*Culicidae*) search for their hosts by detecting smells with their antenna. Their antenna can detect small amounts of exhaled carbon dioxide allowing them to find a target.

Left 100μm   Right 5μm

26

27

Common Housefly
(*Musca domestica*) can live up to a month after emerging from its pupa stage.

400μm

The **Yellow Sac Spider** (*Cheiracanthium mildei*) is a common house spider in the US. They rarely grow larger than 1cm, and their bite is relatively harmless to humans.

400μm

The Fangs of the Yellow Sac Spider (*Cheiracanthium mildei*) have ridges that act like teeth allowing the spider to hold onto its prey.

Left 60μm   Right 5μm

30

31

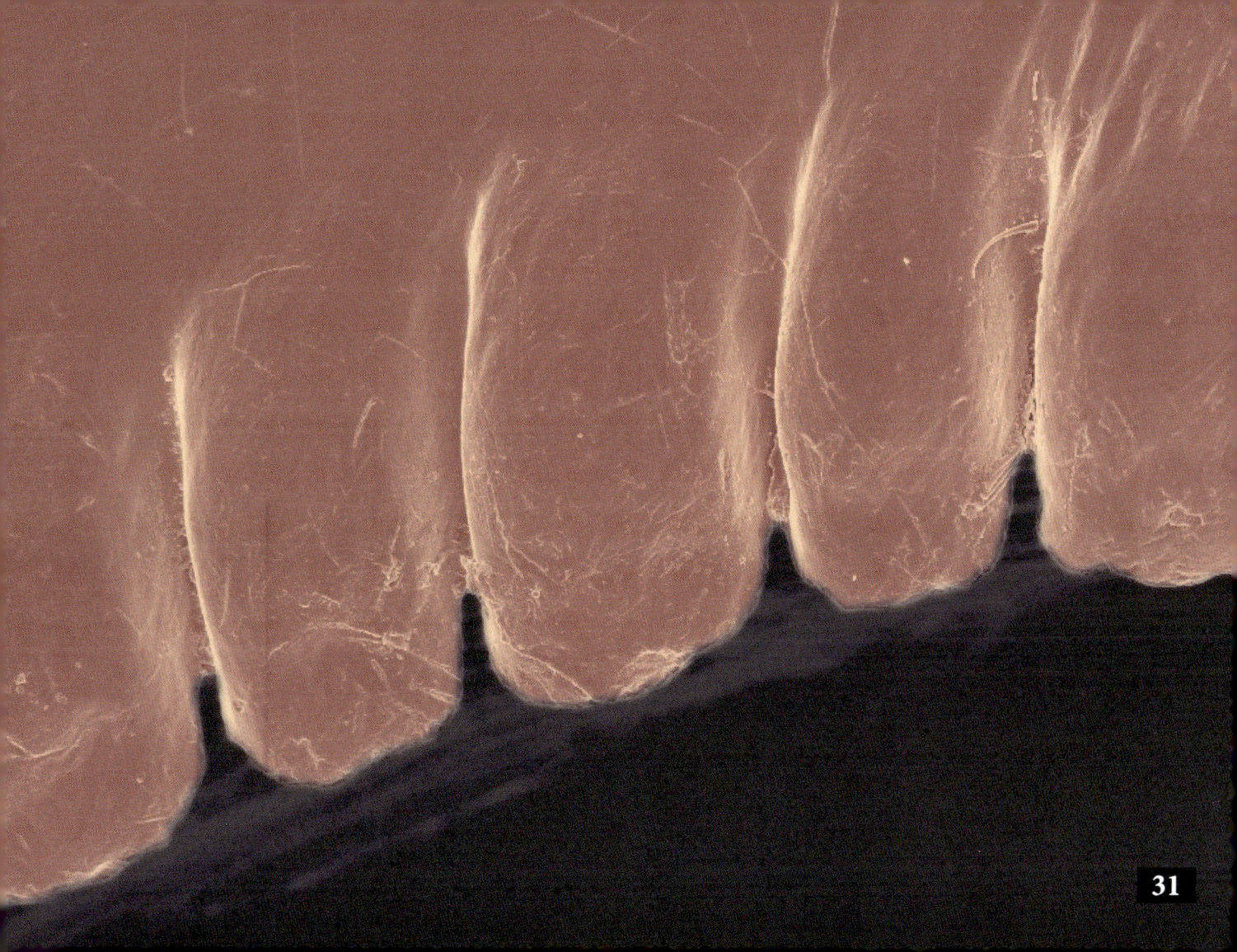

Fireflies (*Lampyridae*) have specialized feet equipped with bristle pads. This increases the contact surface area making it possible for them to climb smooth surfaces like a waxy leaf or glass jar.

50μm

3μm

Crickets (*Gryllidae*) lack the specialized bristle pads of fireflies. While they may not be able to climb smooth surfaces, their spiny hair anatomy makes them more adept at speedy movements required to escape predators.

100μm

House Ants (*Tapinoma sessile*) are too small to pierce human skin. Like most ants they possess a structure at each ankle joint called a spur. Spurs are used like a combs to brush dirt from antennas.

250μm

30μm   20μm   **35**

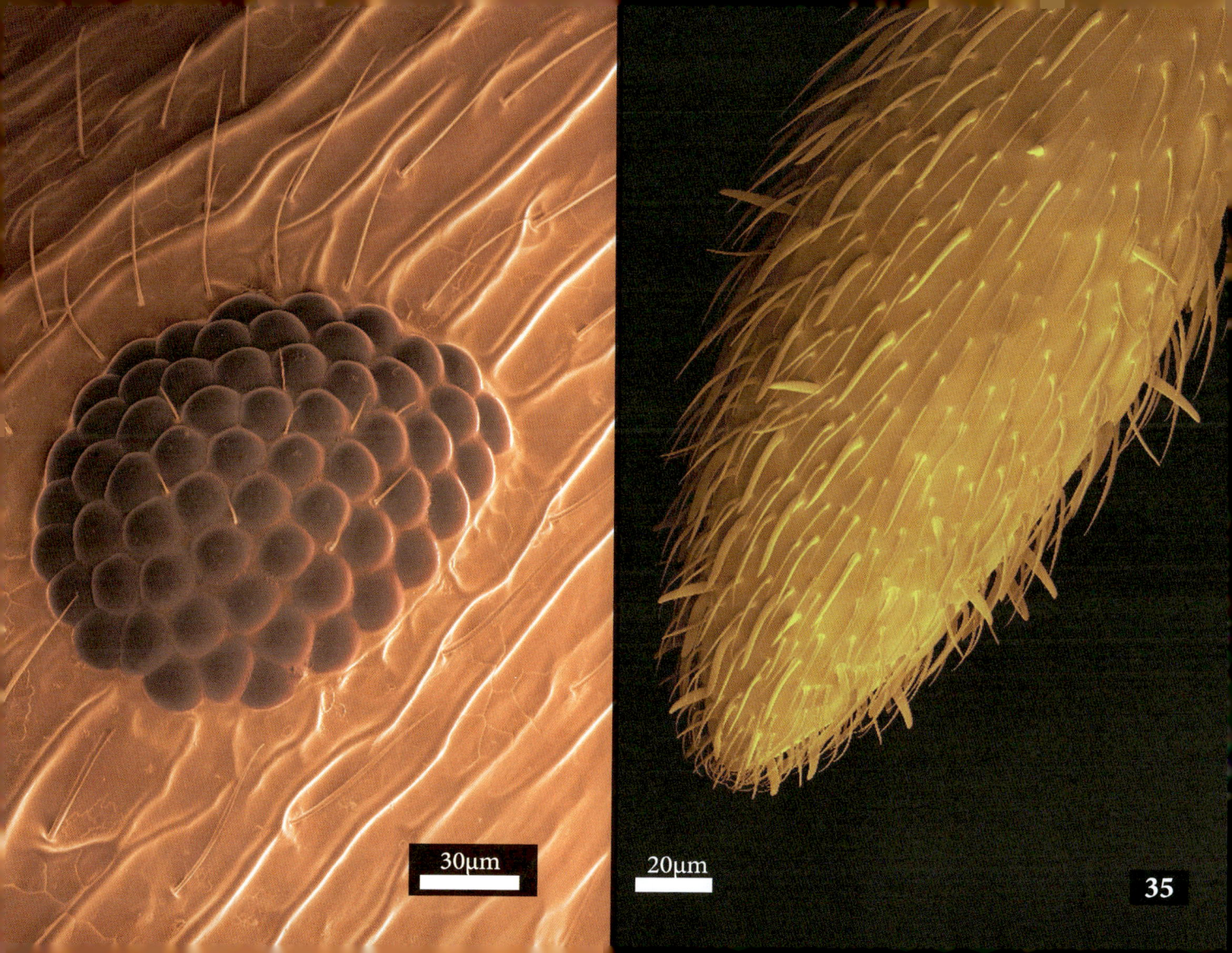

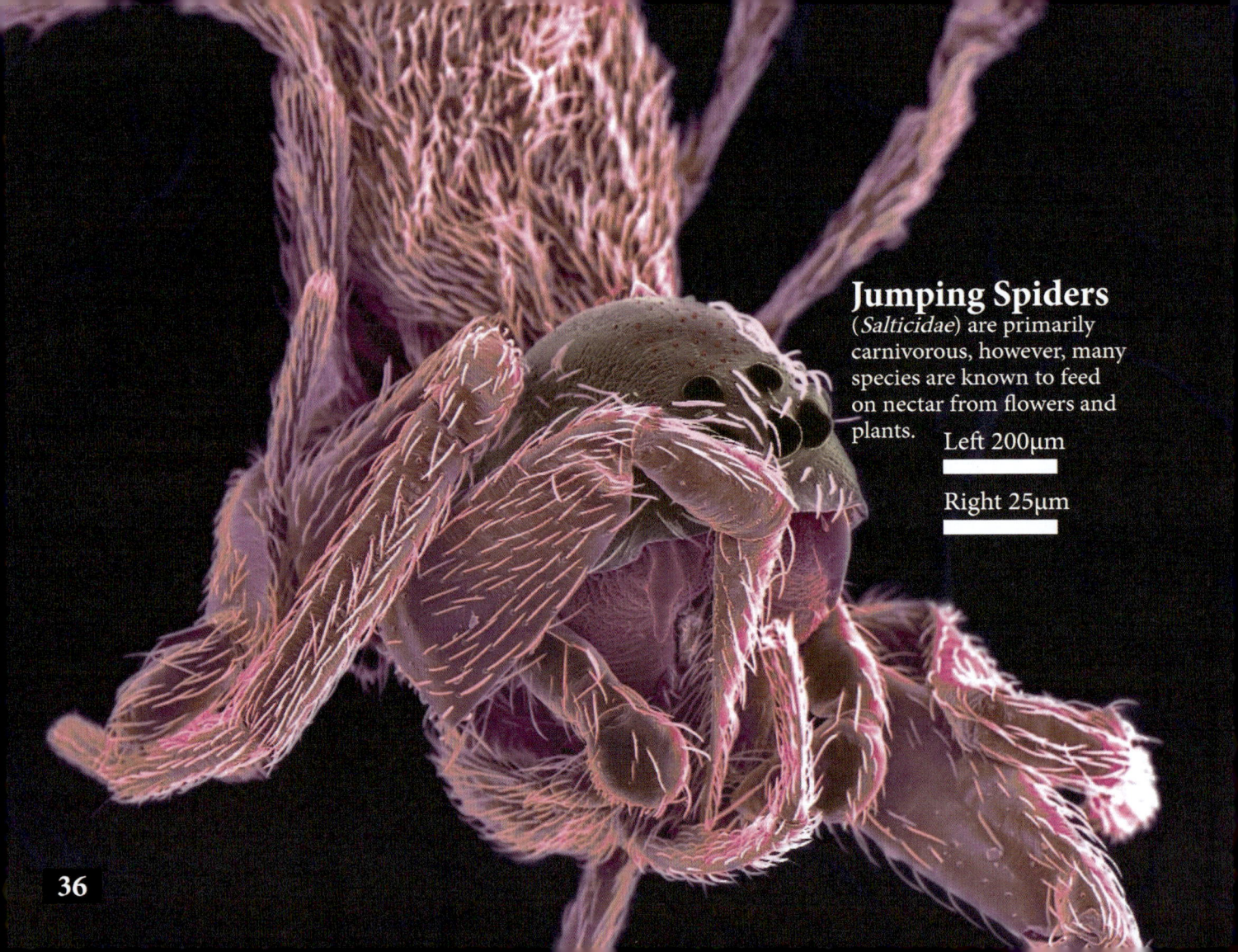

Jumping Spiders (*Salticidae*) are primarily carnivorous, however, many species are known to feed on nectar from flowers and plants.

Left 200μm

Right 25μm

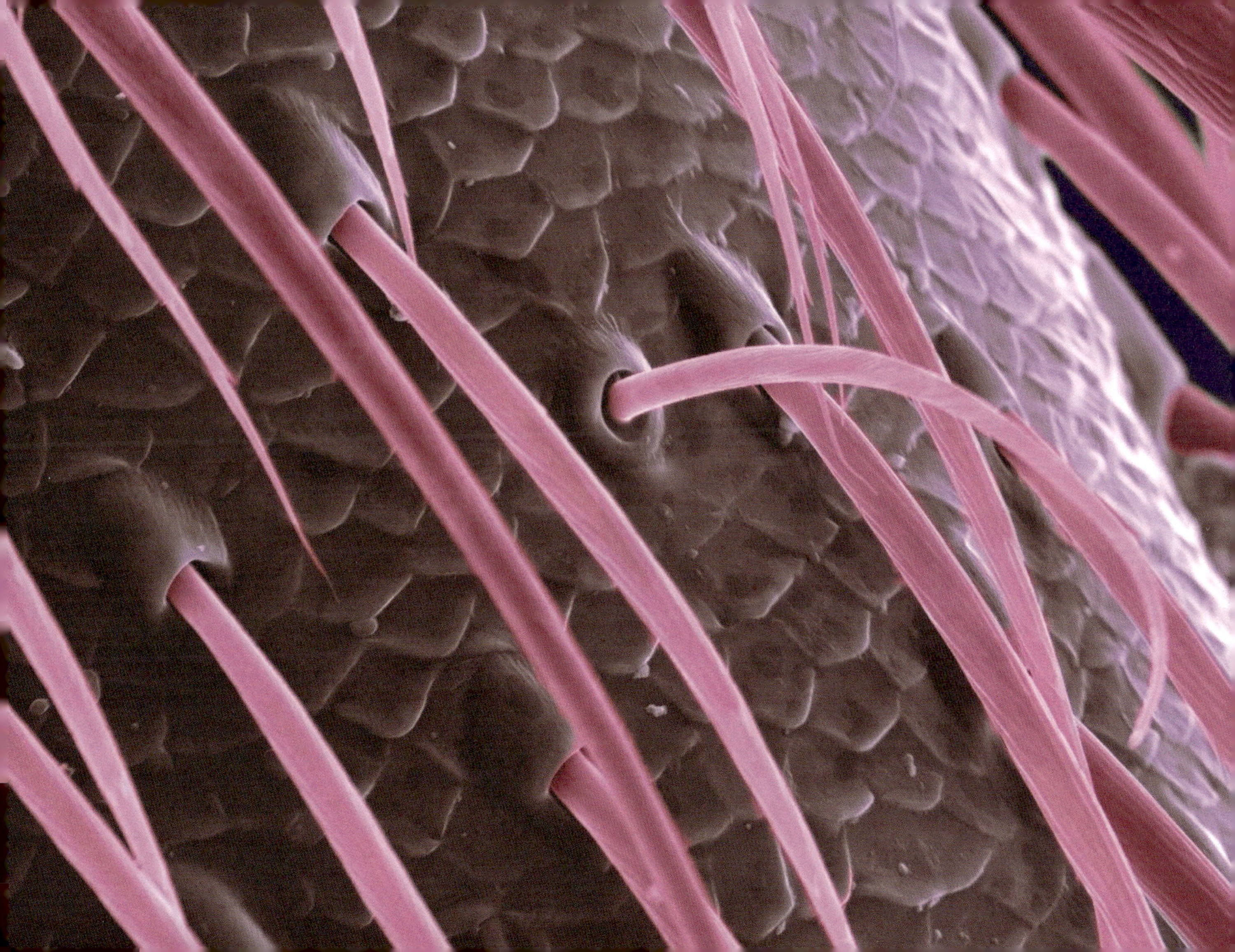

Jumping Spider Continued
Left 15μm   Right 10μm

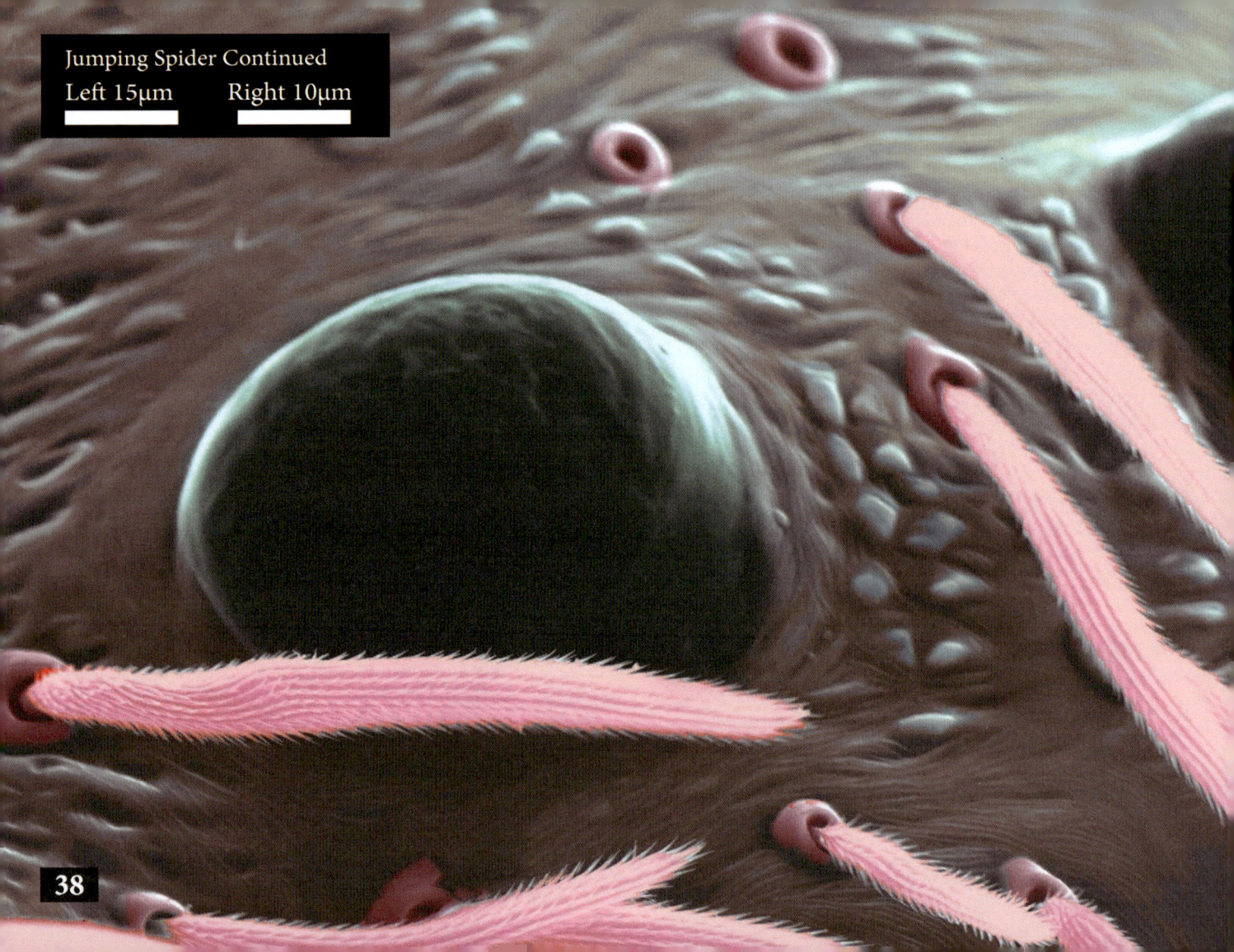

38

39

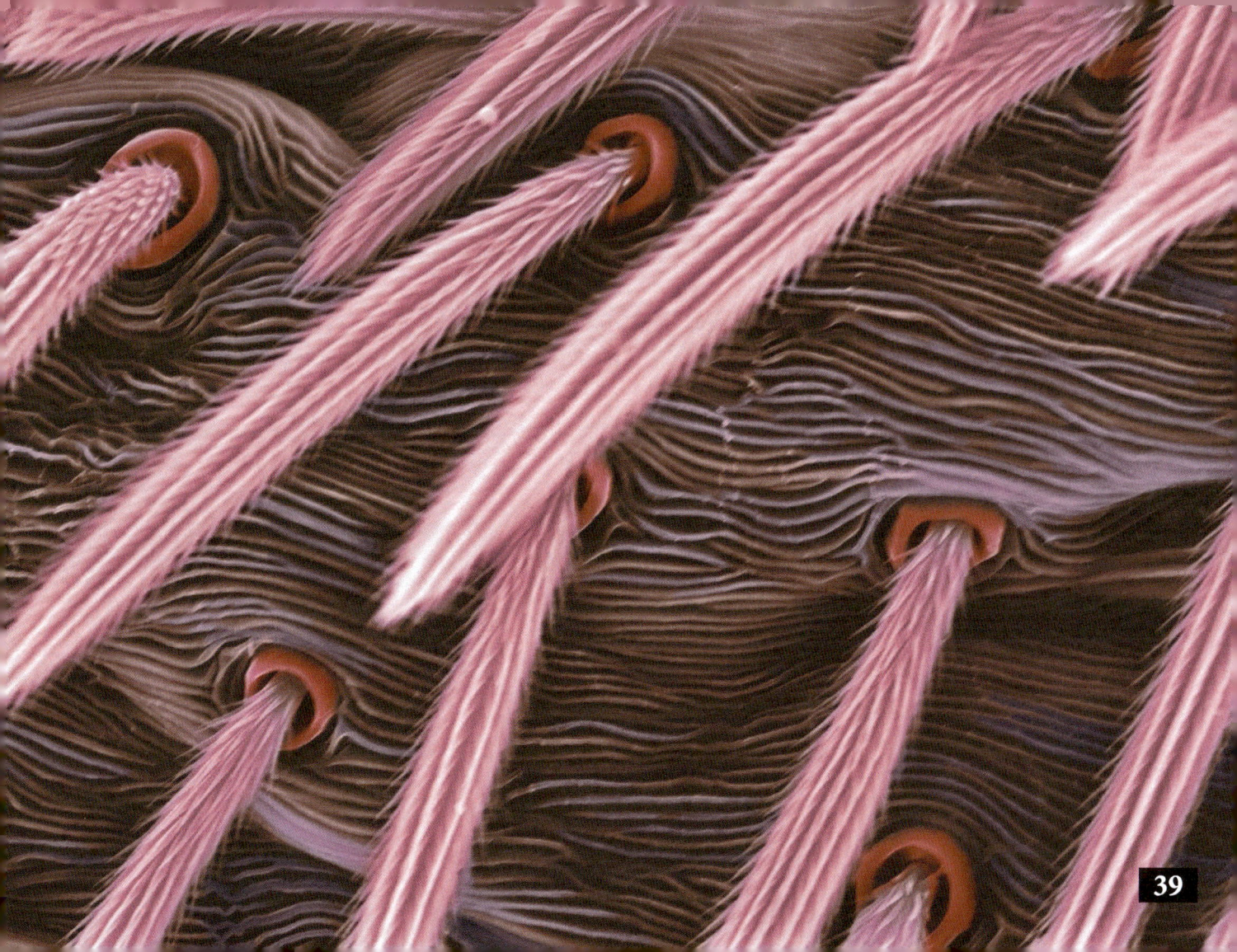

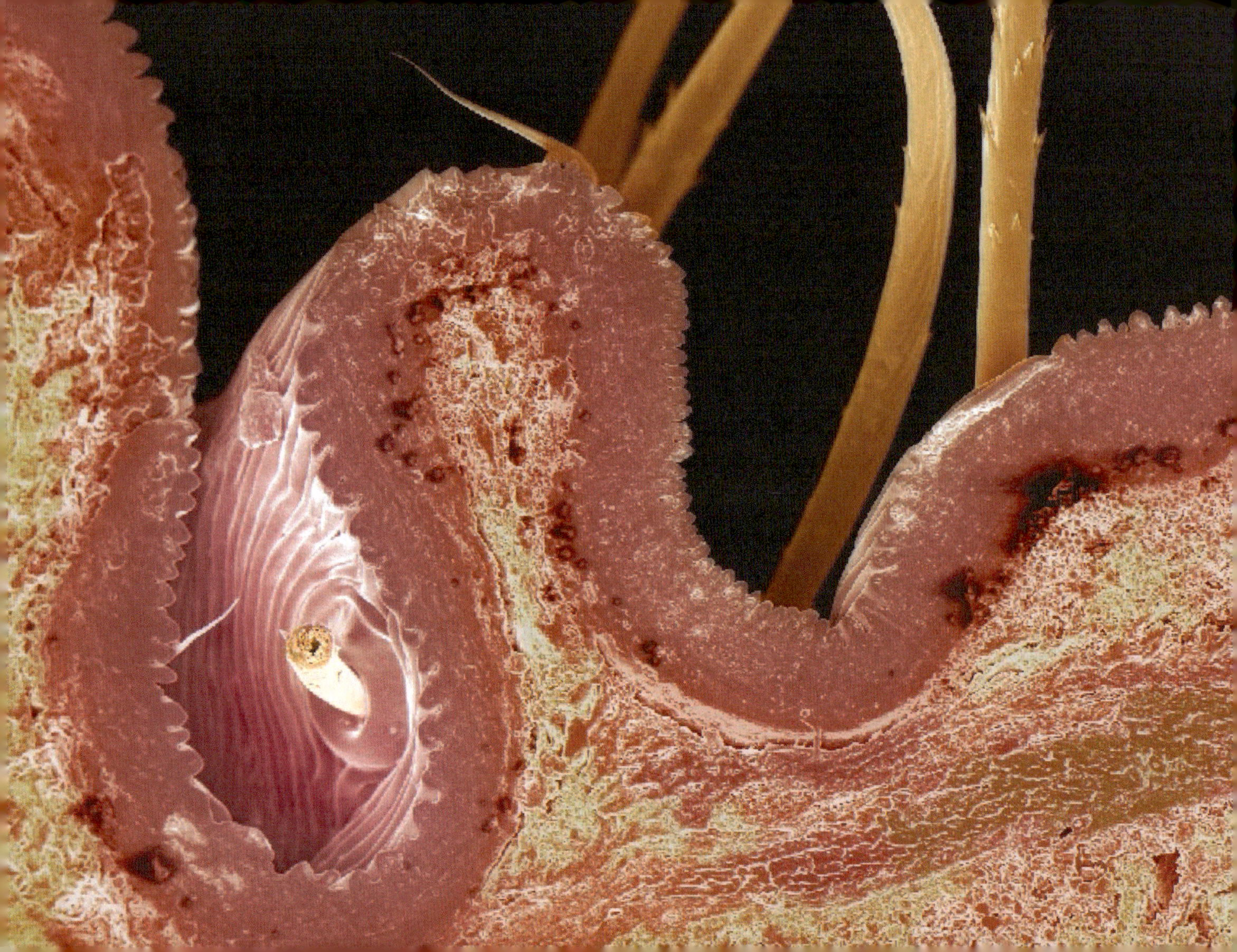

Cross Orbweaver Spiders (*Araneus diadematus*).
Shown here is a cross-sectional view of the spider's hairy abdomen.

10μm

41

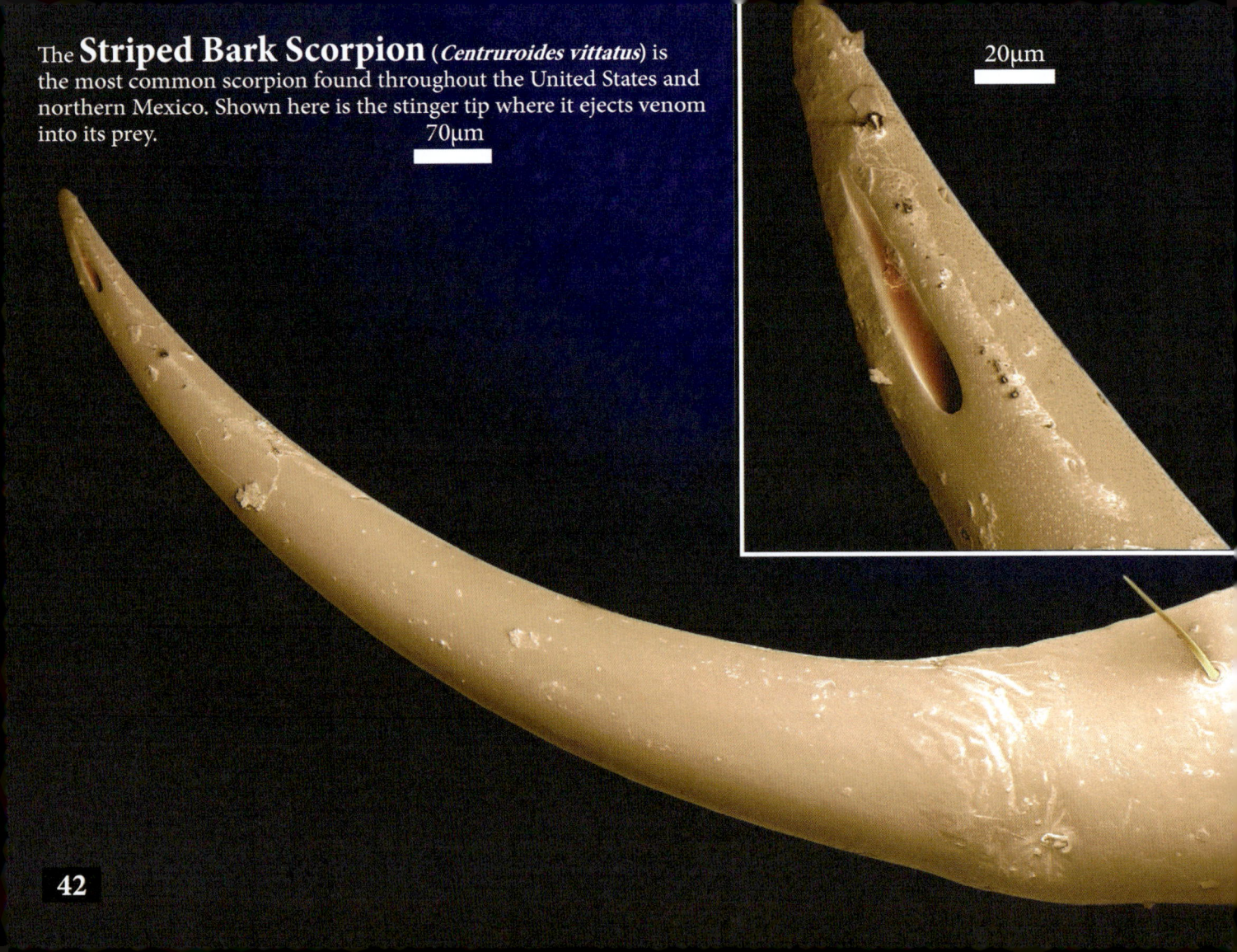

The **Striped Bark Scorpion** (*Centruroides vittatus*) is the most common scorpion found throughout the United States and northern Mexico. Shown here is the stinger tip where it ejects venom into its prey.

70μm

20μm

The **European Paper Wasp** (*Polistes dominula*) is an invasive species in North America. Their yellow and black coloring resembles a hornet. Shown here is the compound eye of the wasp. 10μm

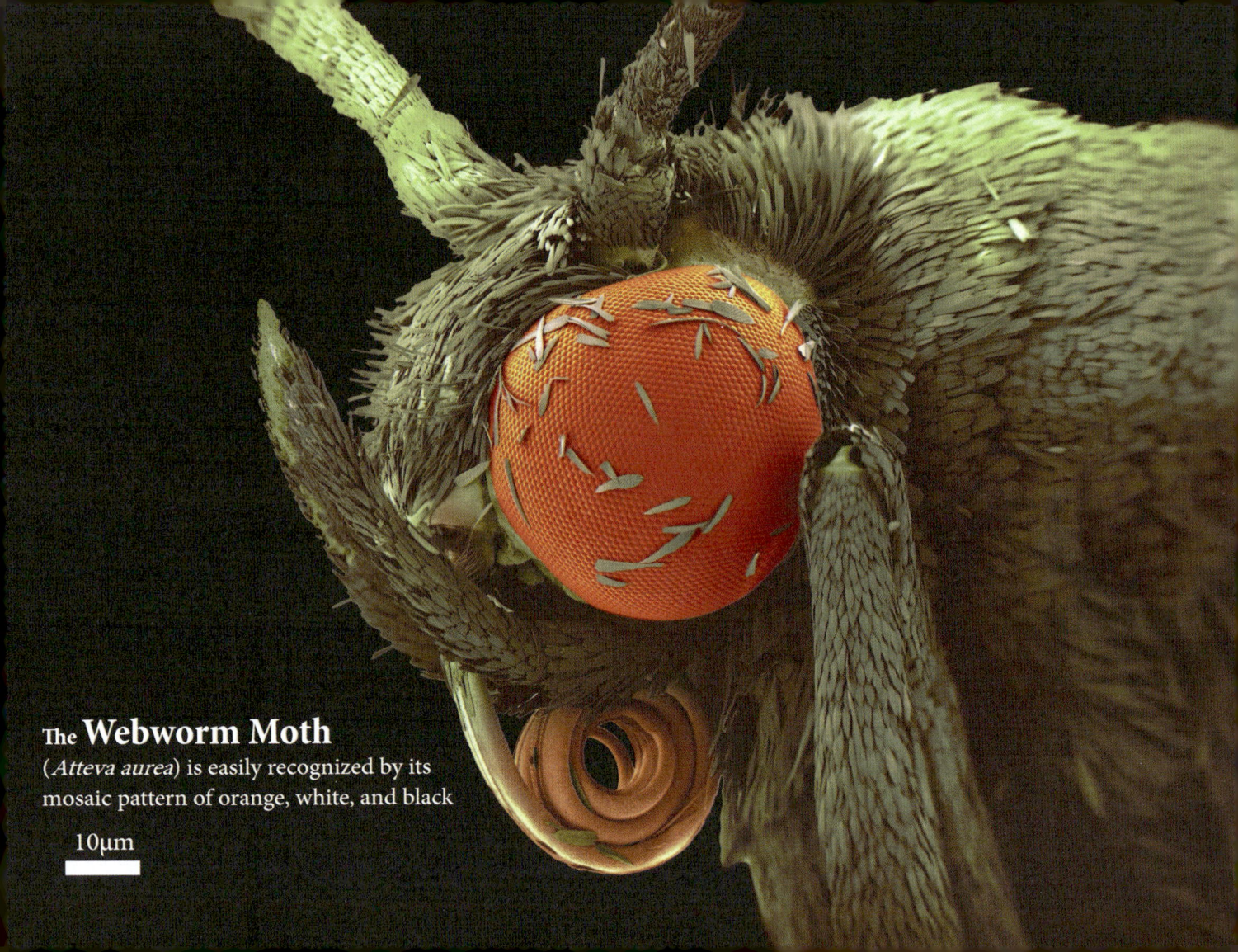

The Webworm Moth
(*Atteva aurea*) is easily recognized by its mosaic pattern of orange, white, and black

10μm

Honey Bees (*Apis mellifera*) use their long tongues (proboscis) to reach nectar deep within flowers.

10μm

CREATURES

Like most amphibians, the **Eastern Newt** (*Notophthalmus viridescens*) has tiny pores in its skin. These pores allow the newt to breathe through its skin by diffusion of oxygen and carbon dioxide. Using this method of cutaneous respiration some amphibians can spend hours underwater without coming up for air.

1μm

Earthworms (*Lumbricus terrestris*) move through the ground using waves of muscle contractions called peristalsis. Their body is covered in claw-like bristles (setae) that provide traction during movement.

100μm

49

Leopard Geckos (*Eublepharis macularius*) 125μm
shed their skin about once a month. Shown here is
one of those sheddings.

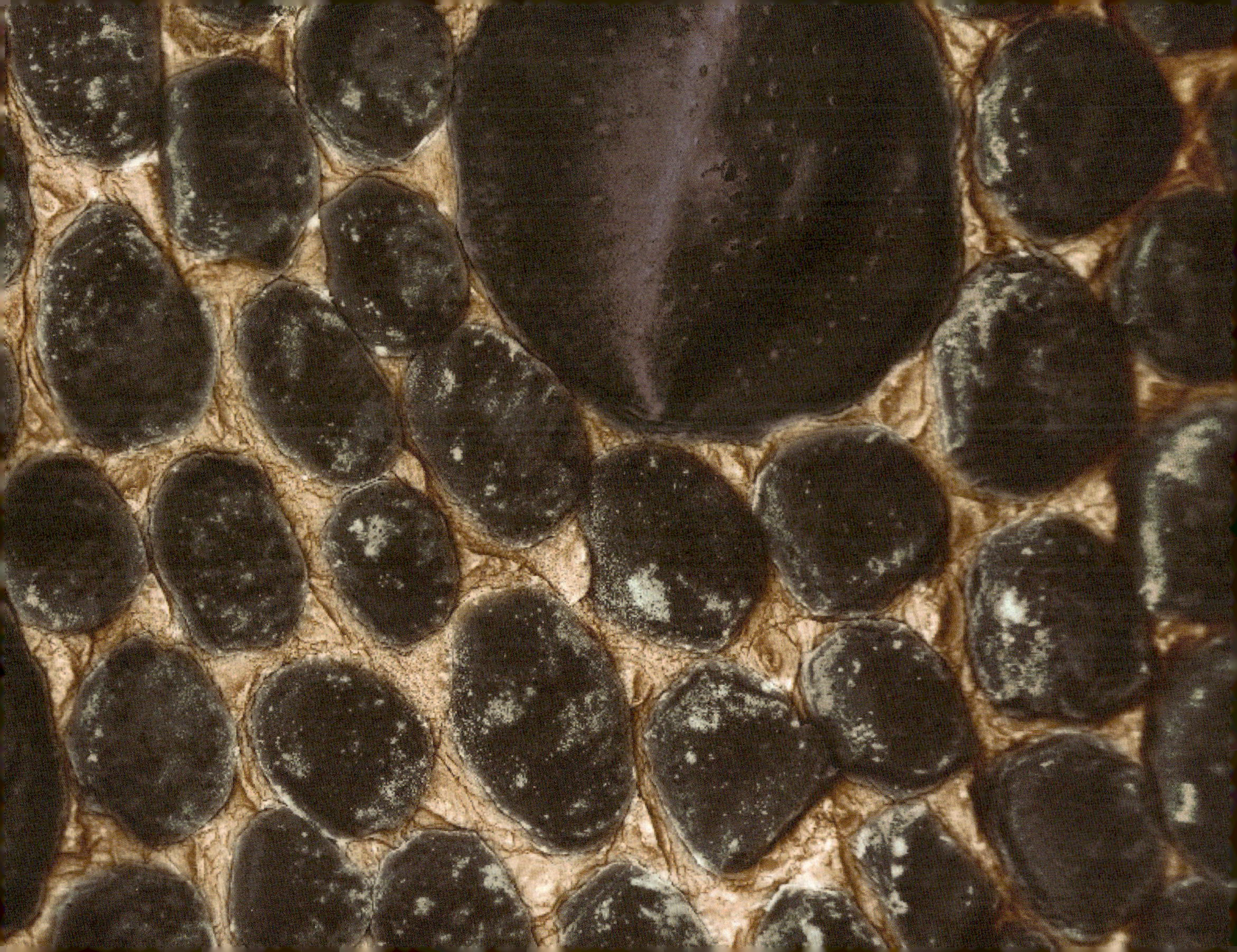

Animal Hair

15μm

Human Arm Hair (Caucasian male)

Cat (Main Coon)

Dog (Golden Doodle)

Rabbit (New Zealand White)

Mouse (Albino)

Mucus (*a.k.a. boogies*) is made up of very large molecules called mucins which are glycoproteins. On a nano scale they form an intertangled web of hydrated fibers.

800nm

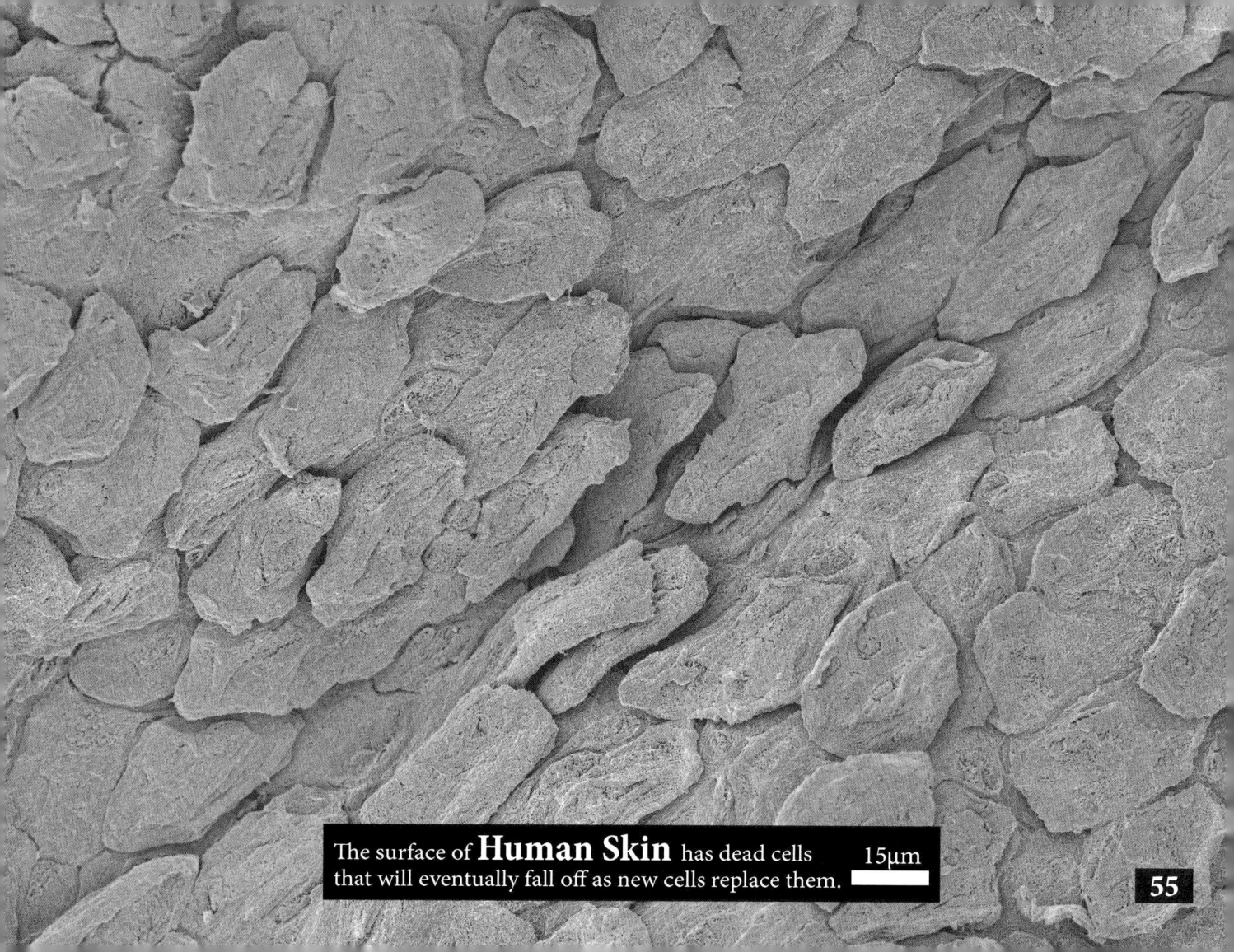

The surface of **Human Skin** has dead cells that will eventually fall off as new cells replace them.

15μm

55

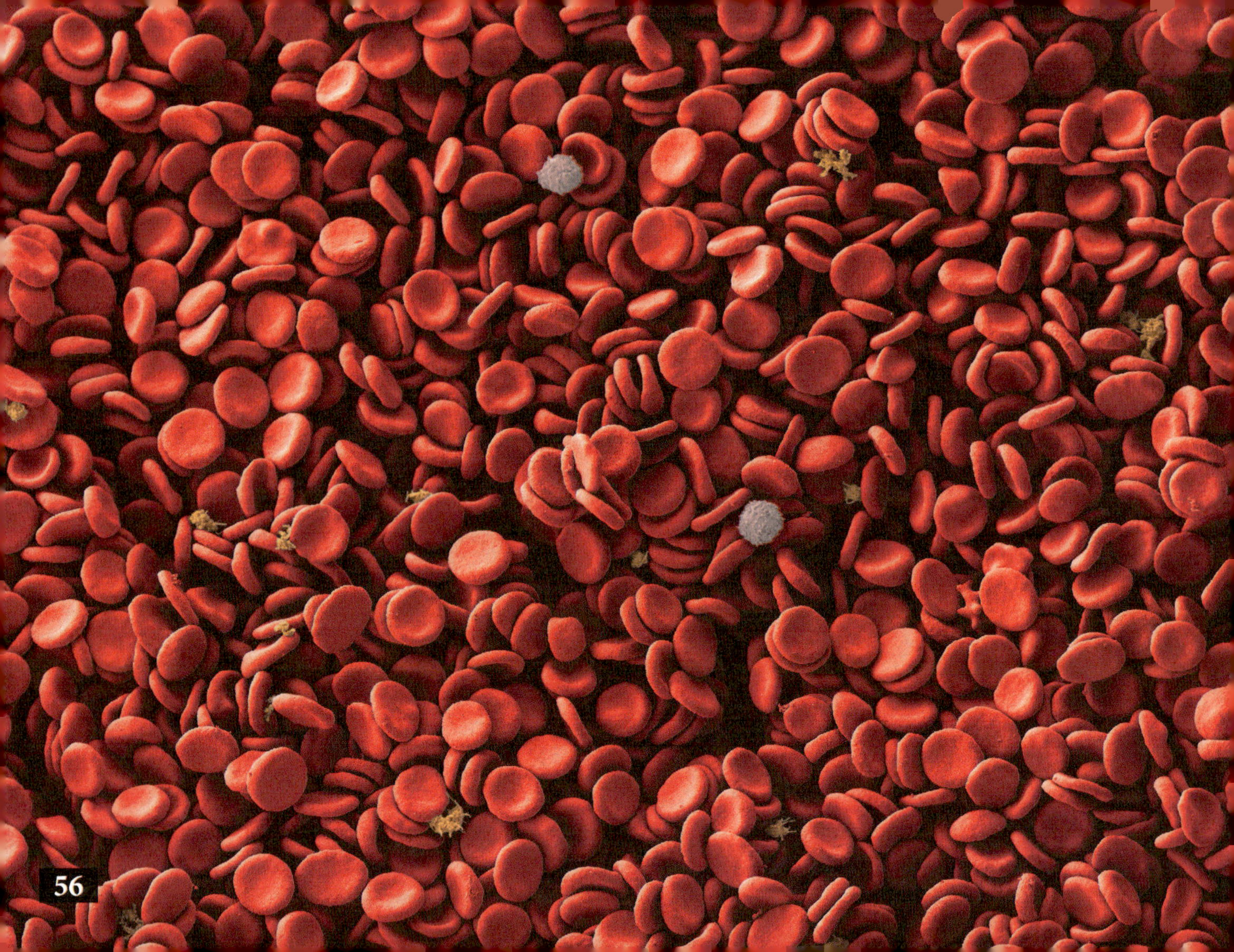

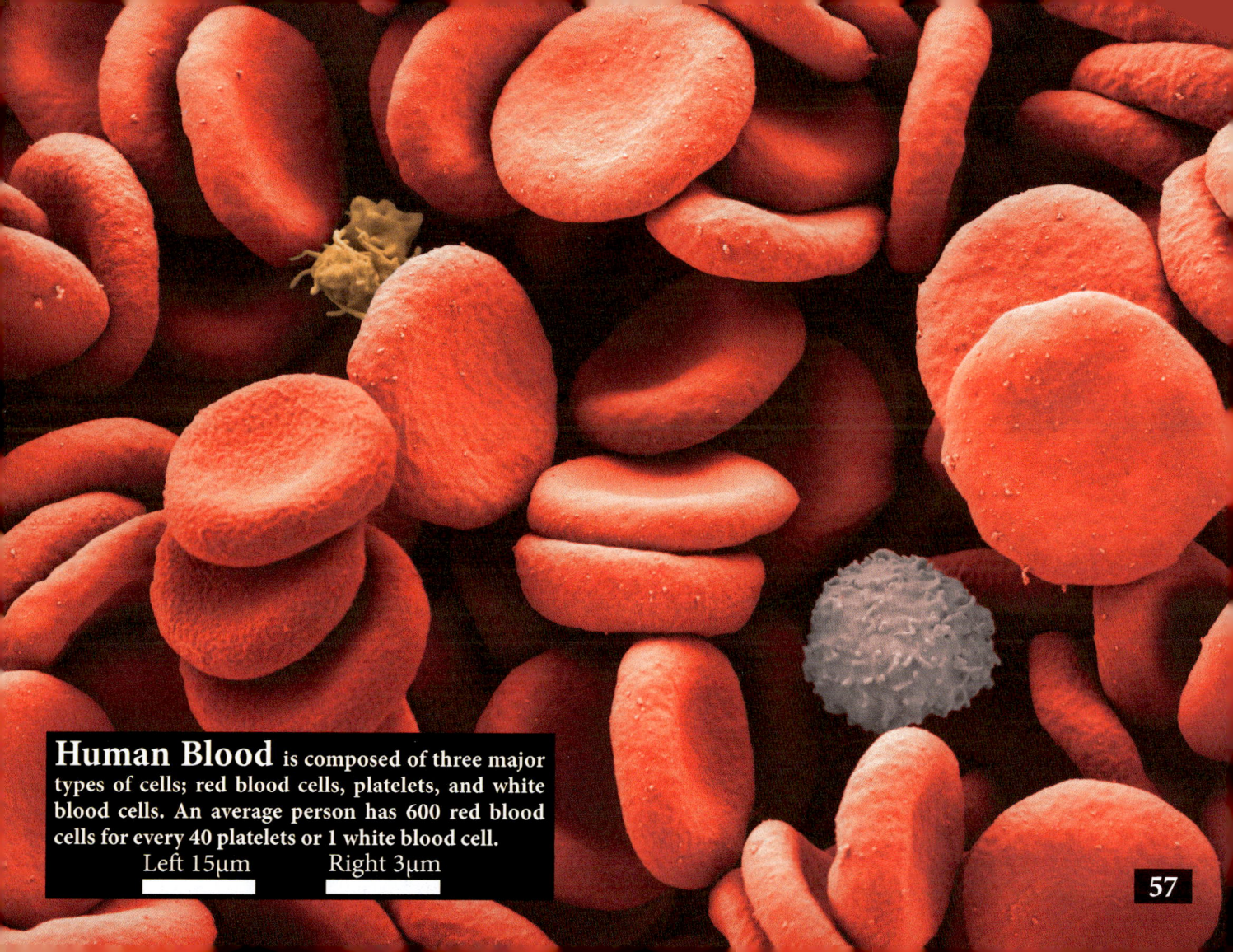

Human Blood is composed of three major types of cells; red blood cells, platelets, and white blood cells. An average person has 600 red blood cells for every 40 platelets or 1 white blood cell.

Left 15µm   Right 3µm

57

COMMON ITEMS

200μm

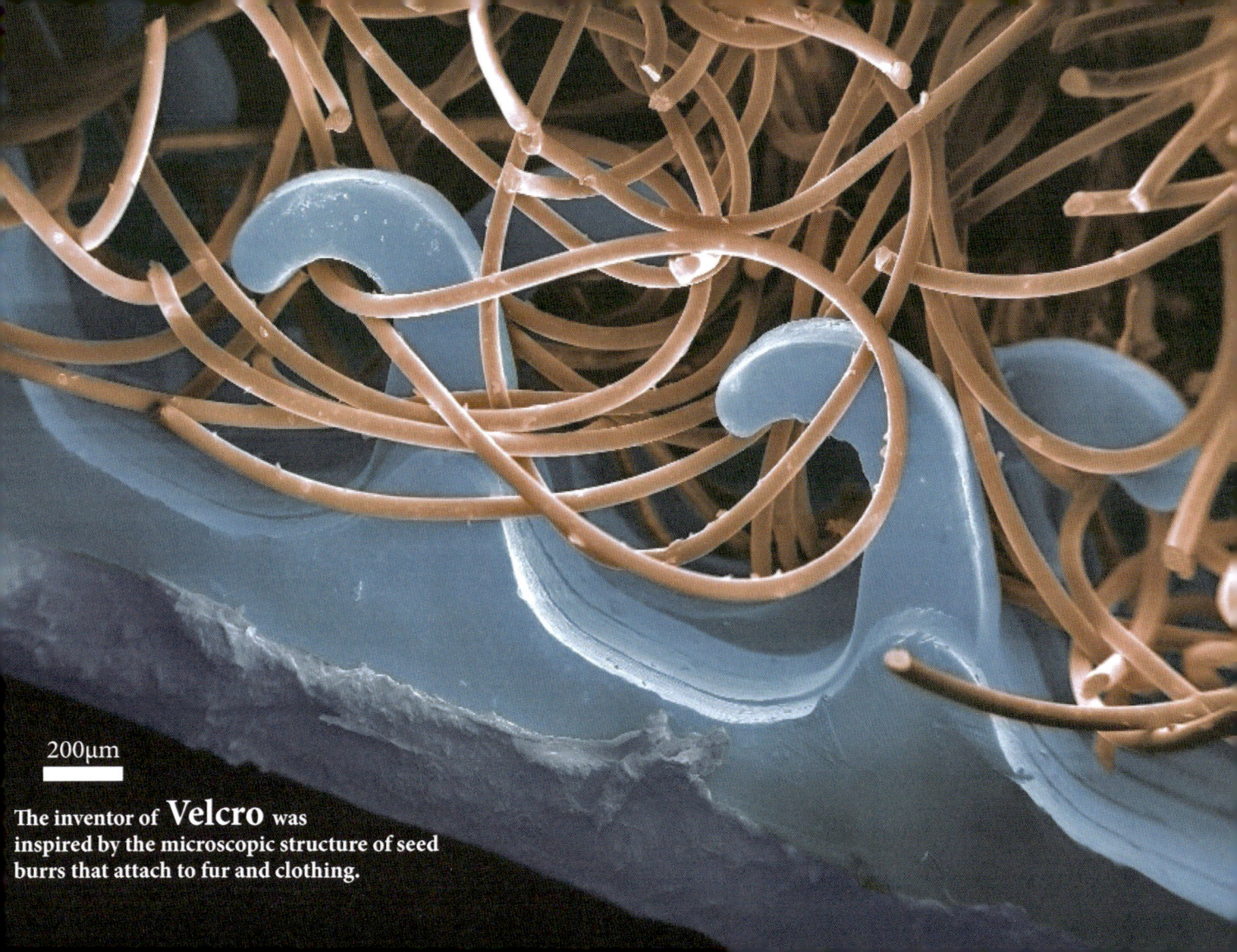

The inventor of **Velcro** was inspired by the microscopic structure of seed burrs that attach to fur and clothing.

Tungsten Light Bulb Filament
Tungsten has the highest melting point of all elements, allowing it to glow white hot before reaching its melting temperature of 3422 °C (6192 °F).

100μm

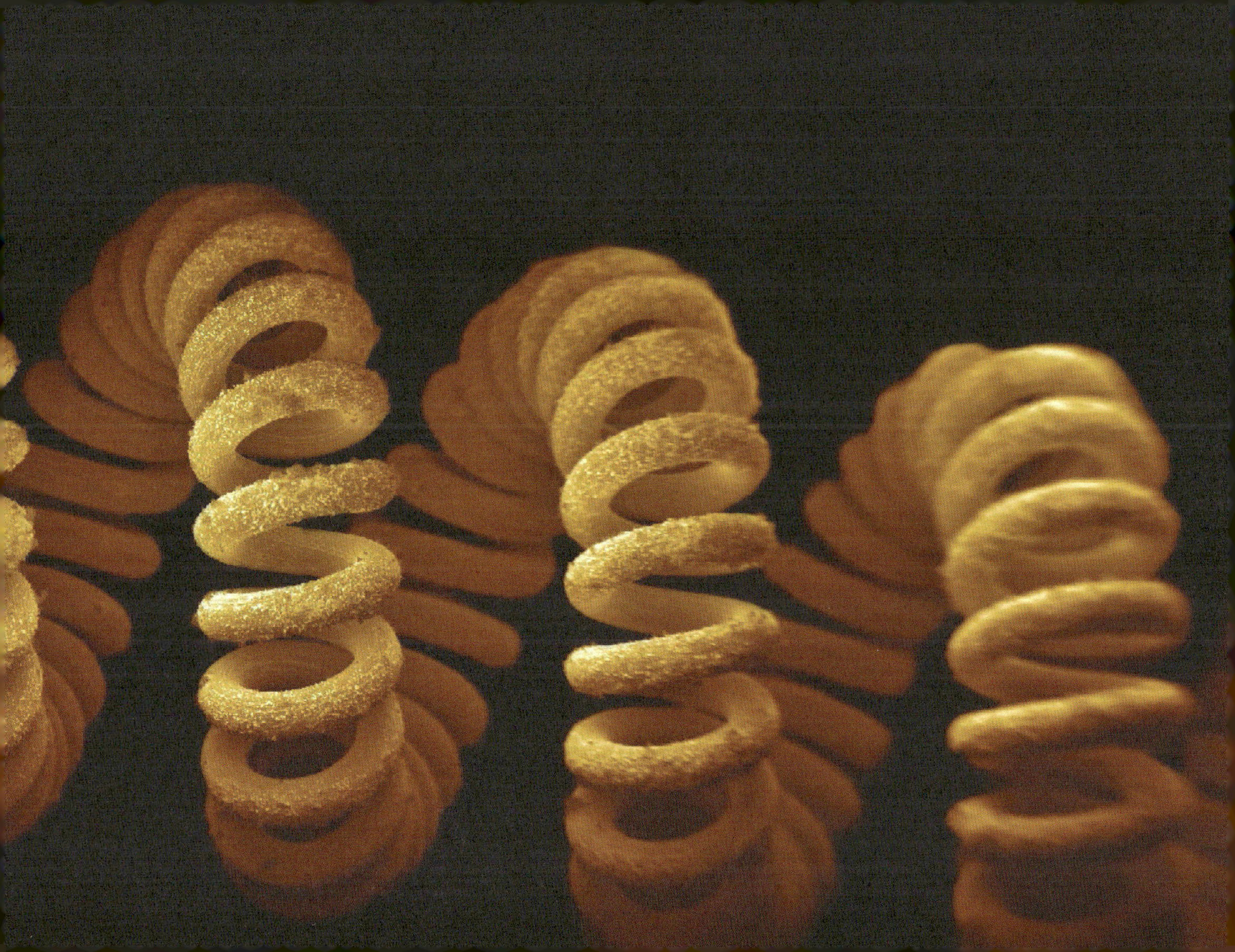

62 **New Toothbrush** 100μm

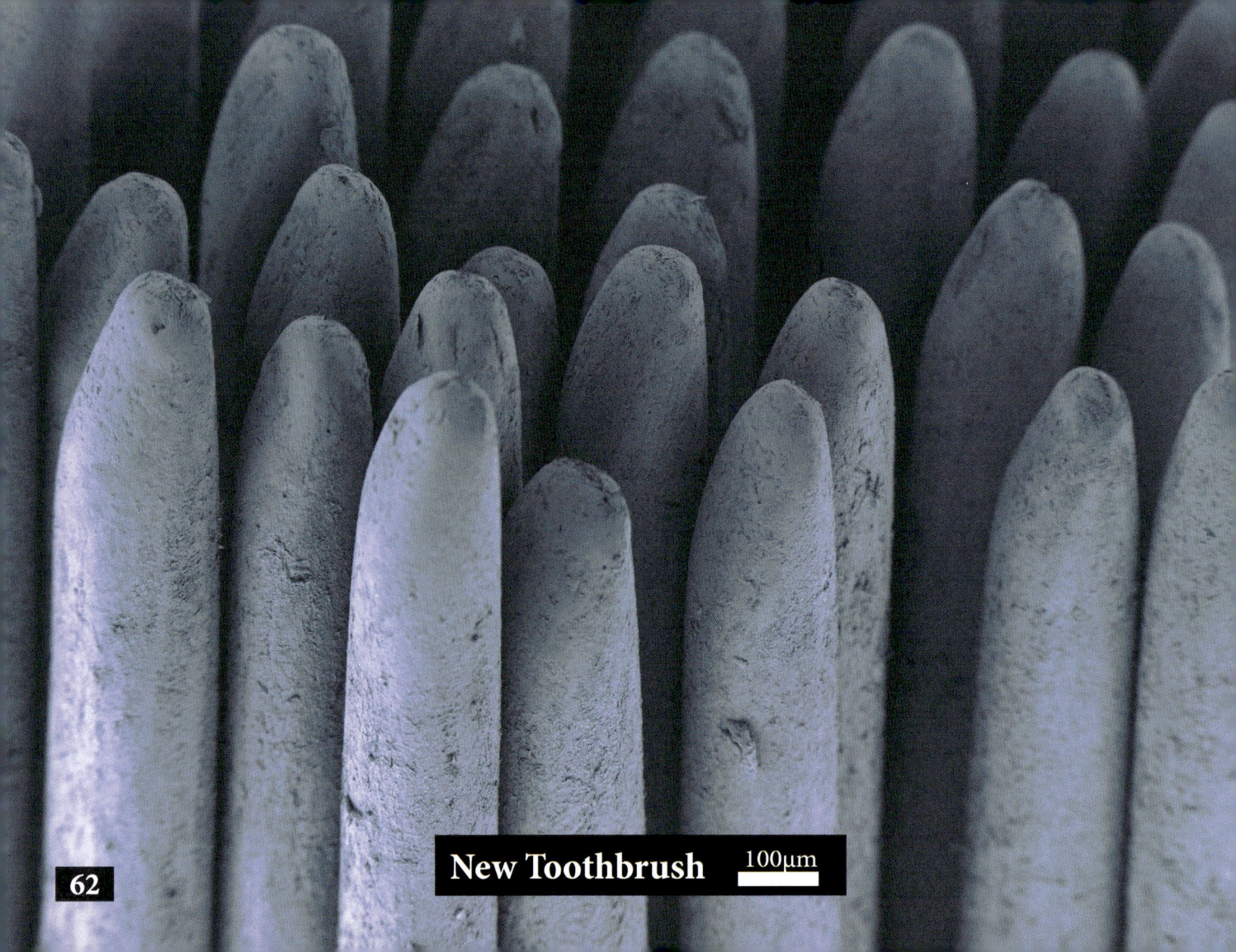

3 Month Old Toothbrush 100μm

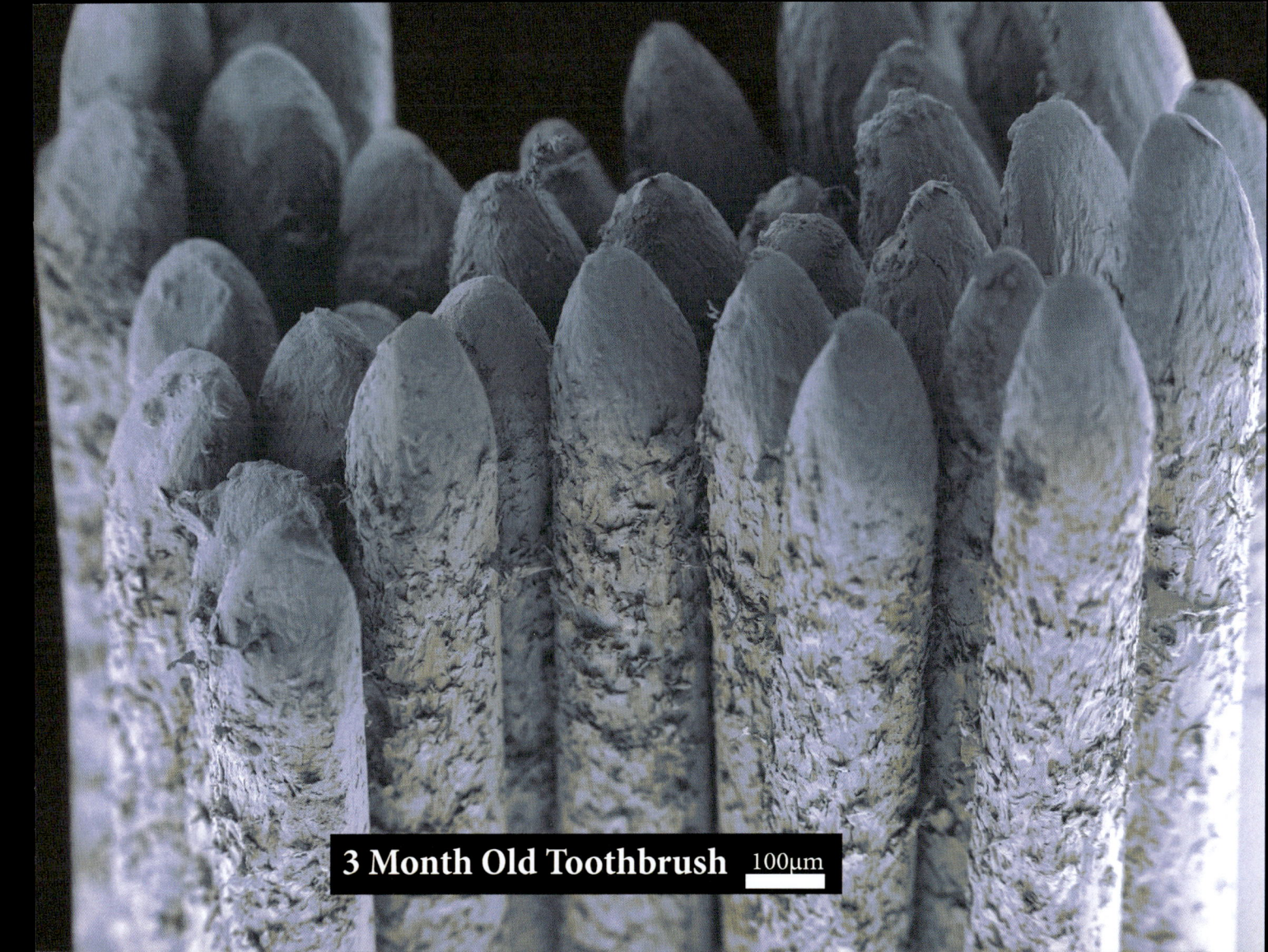

64

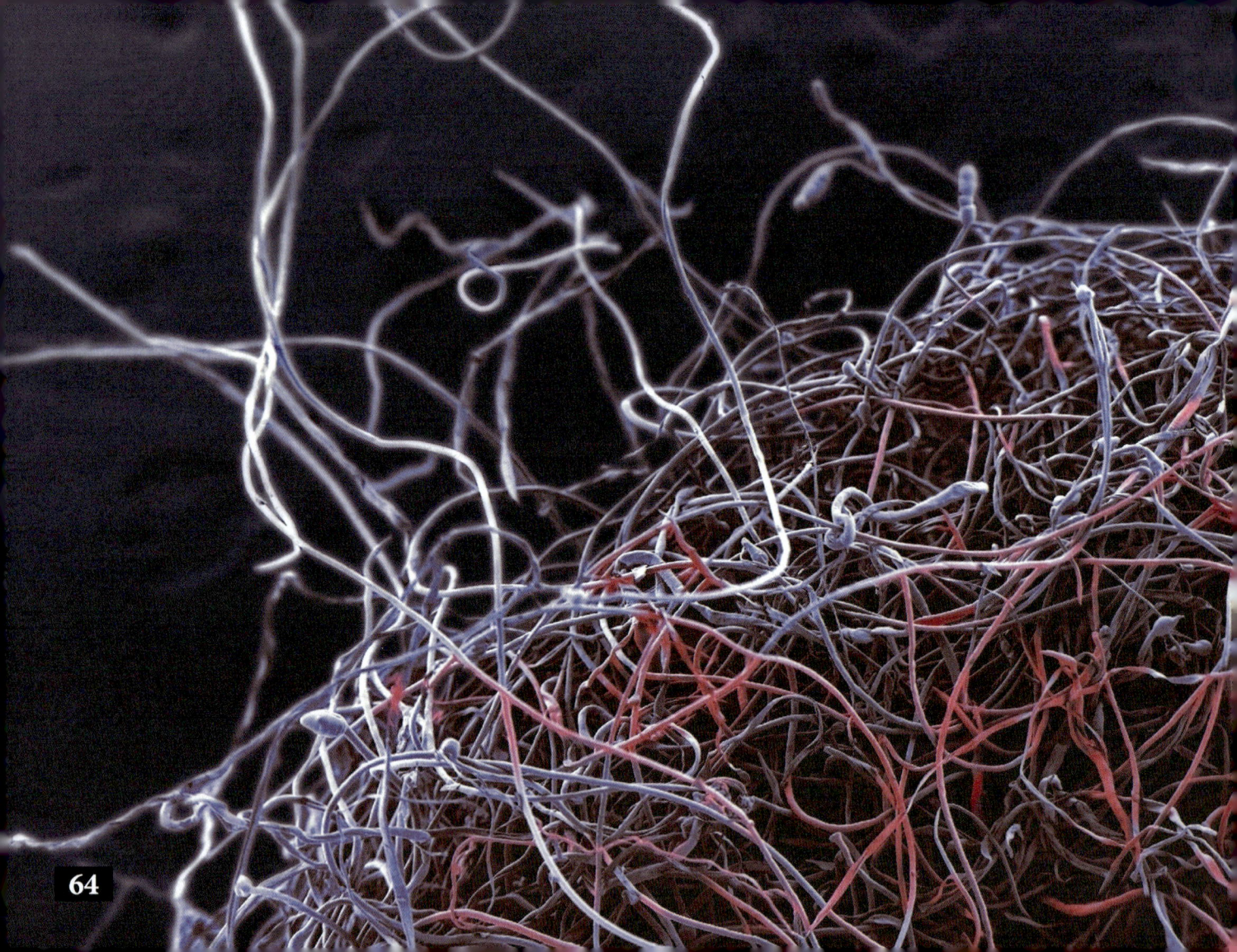

Pocket Lint   150μm

Guitar String 100μm

Polyester Shirt 75μm

BACTERIA, FUNGUS & PLANTS

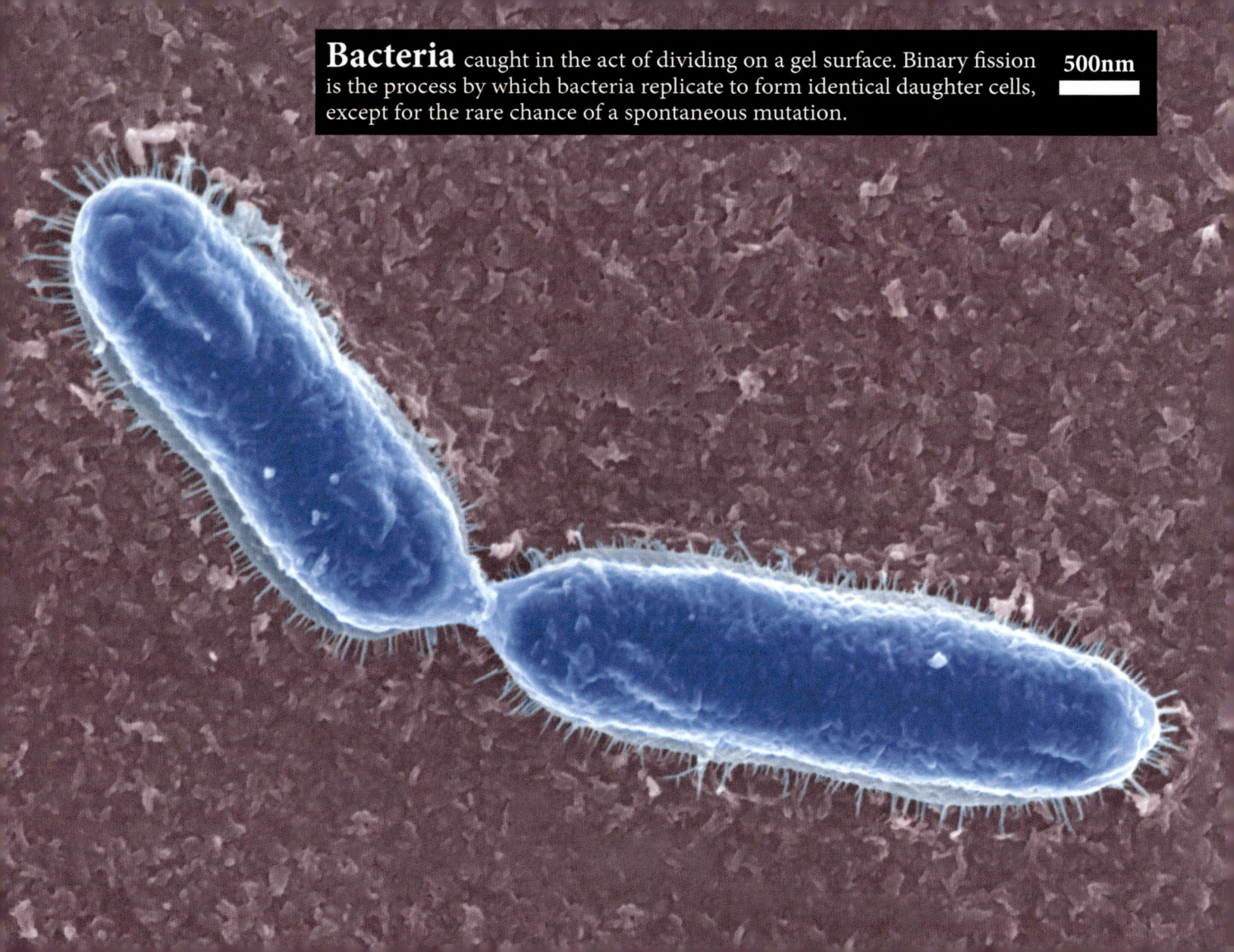

Bacteria caught in the act of dividing on a gel surface. Binary fission is the process by which bacteria replicate to form identical daughter cells, except for the rare chance of a spontaneous mutation.

500nm

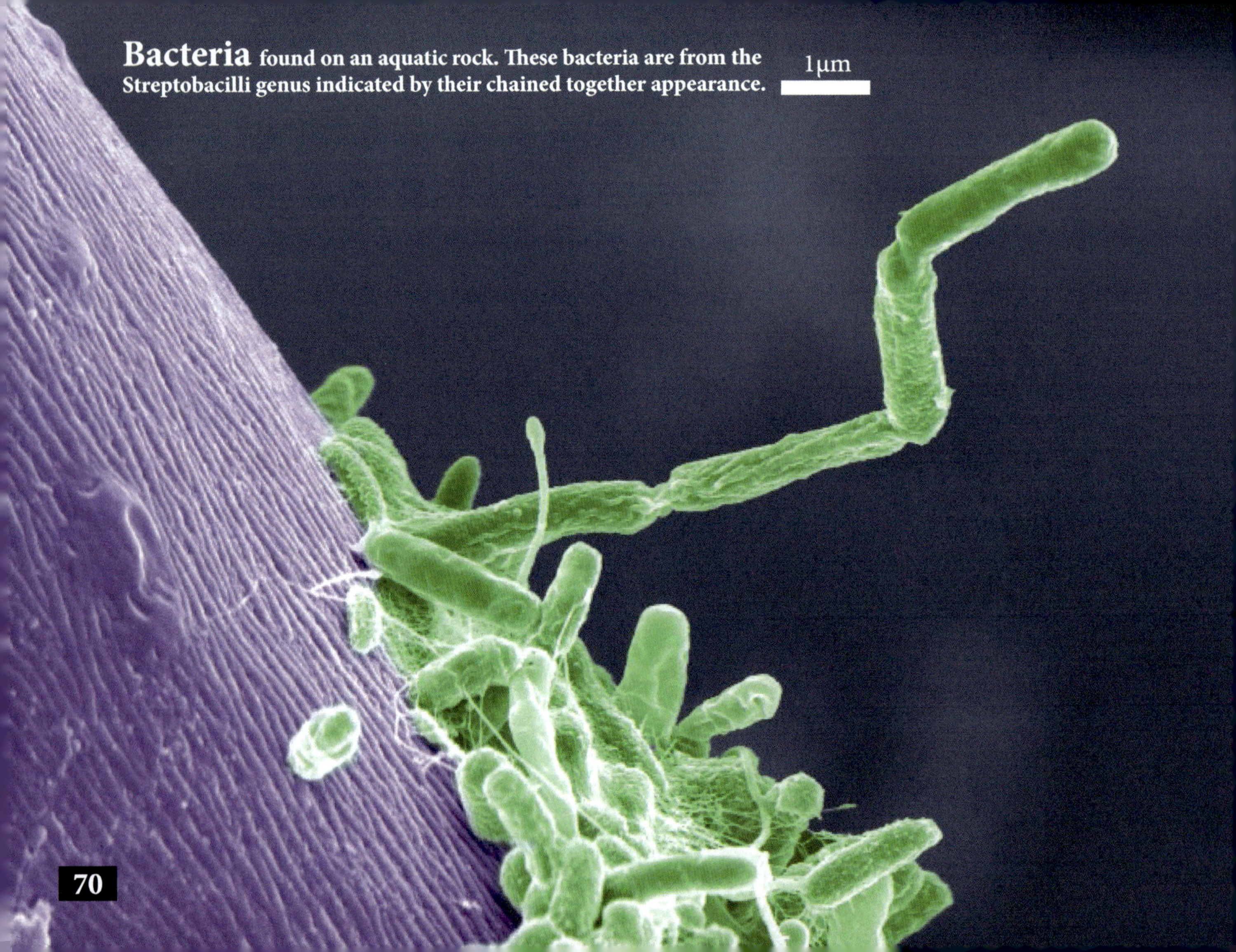

Bacteria found on an aquatic rock. These bacteria are from the Streptobacilli genus indicated by their chained together appearance.

1μm

70

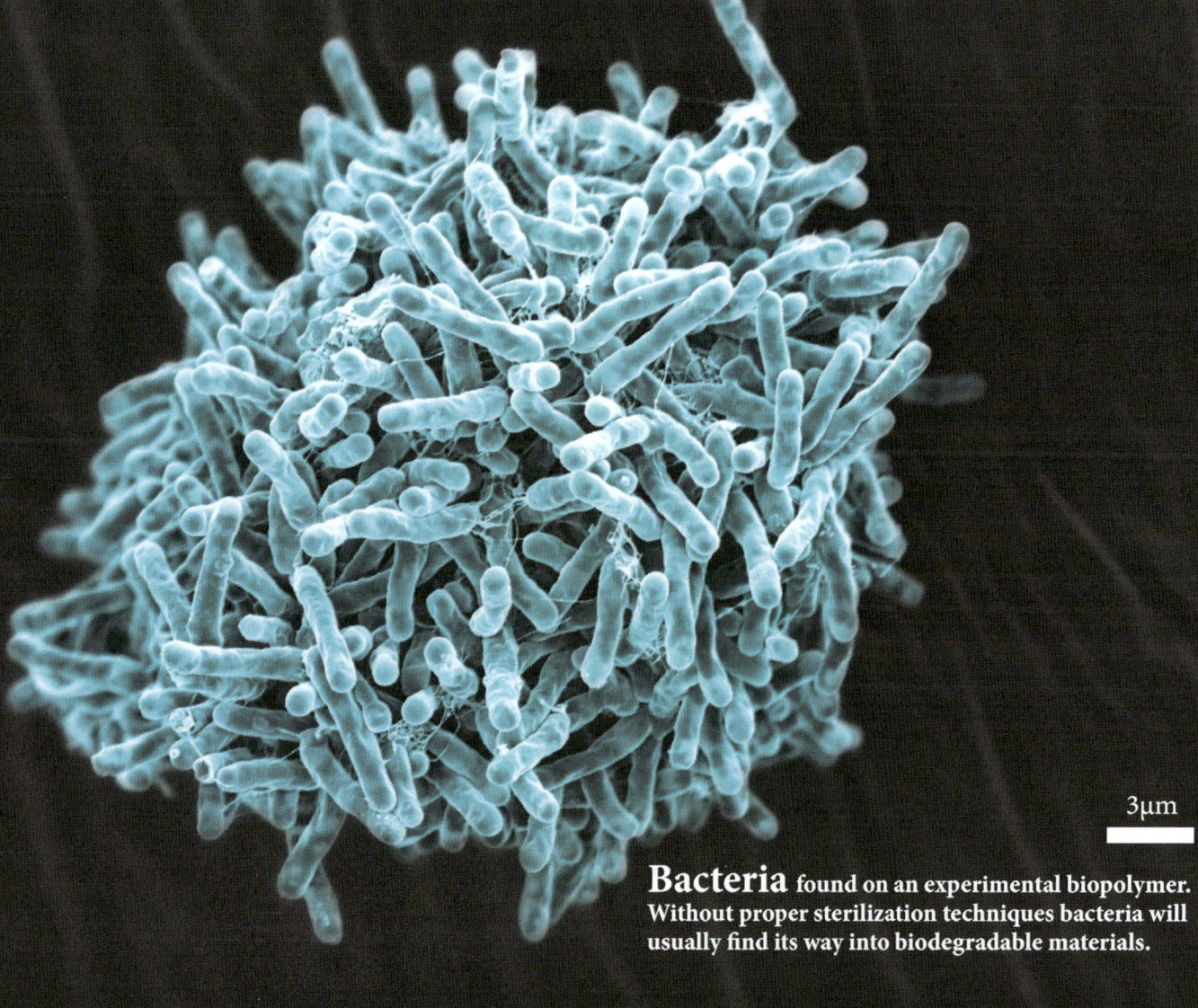

3μm

Bacteria found on an experimental biopolymer. Without proper sterilization techniques bacteria will usually find its way into biodegradable materials.

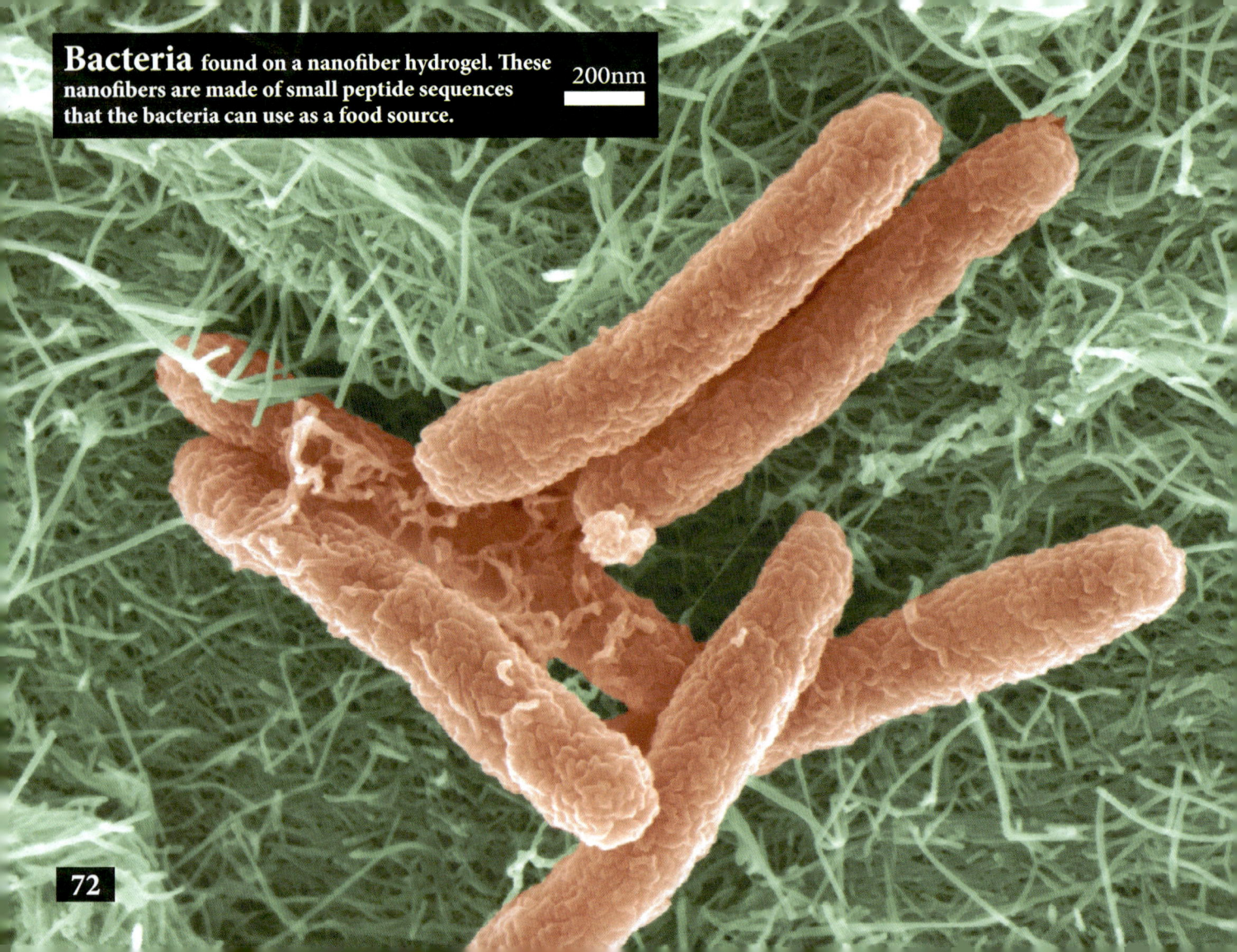

Bacteria found on a nanofiber hydrogel. These nanofibers are made of small peptide sequences that the bacteria can use as a food source.

200nm

72

Algae Cells (diatoms) from a freshwater fish aquarium.   2μm

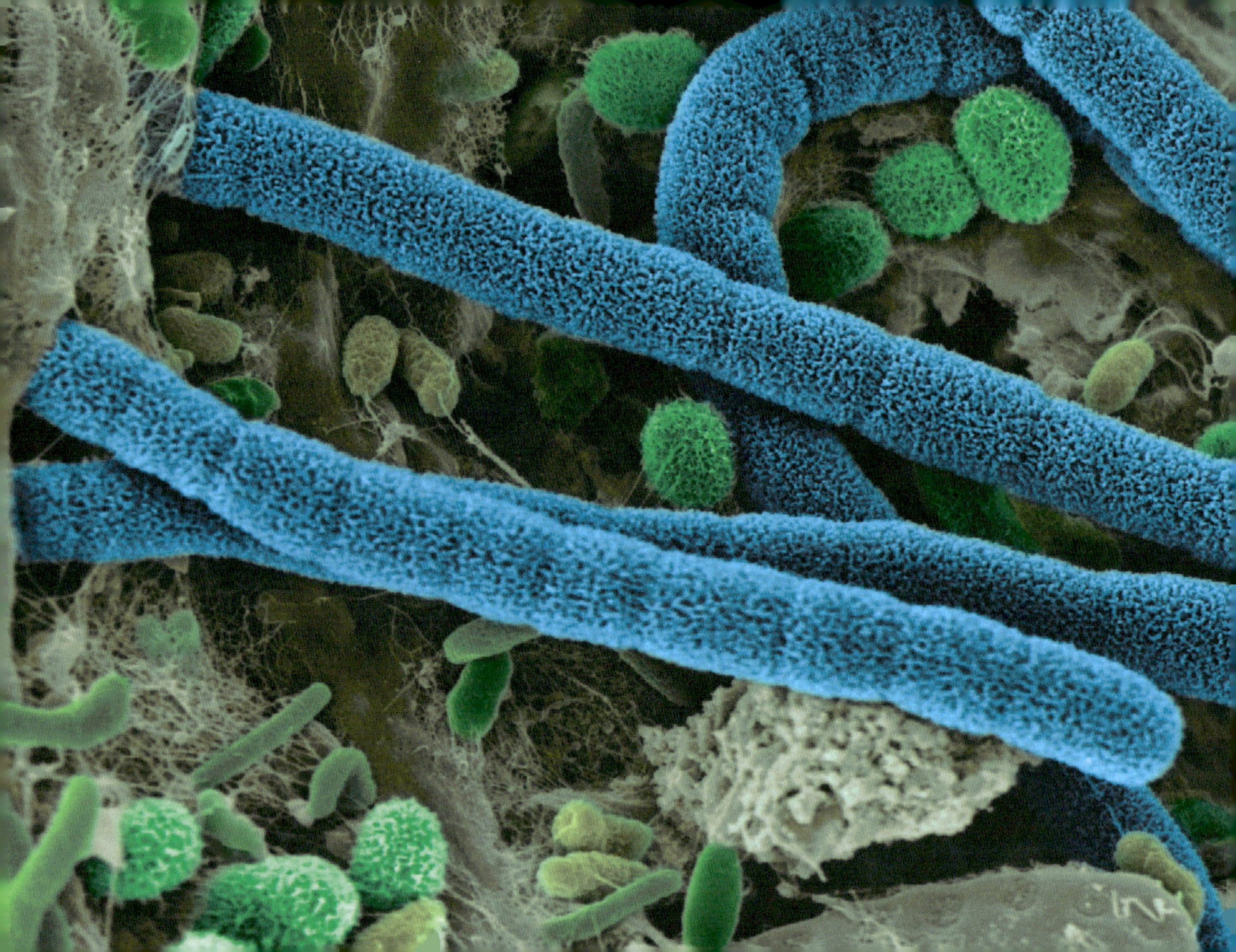

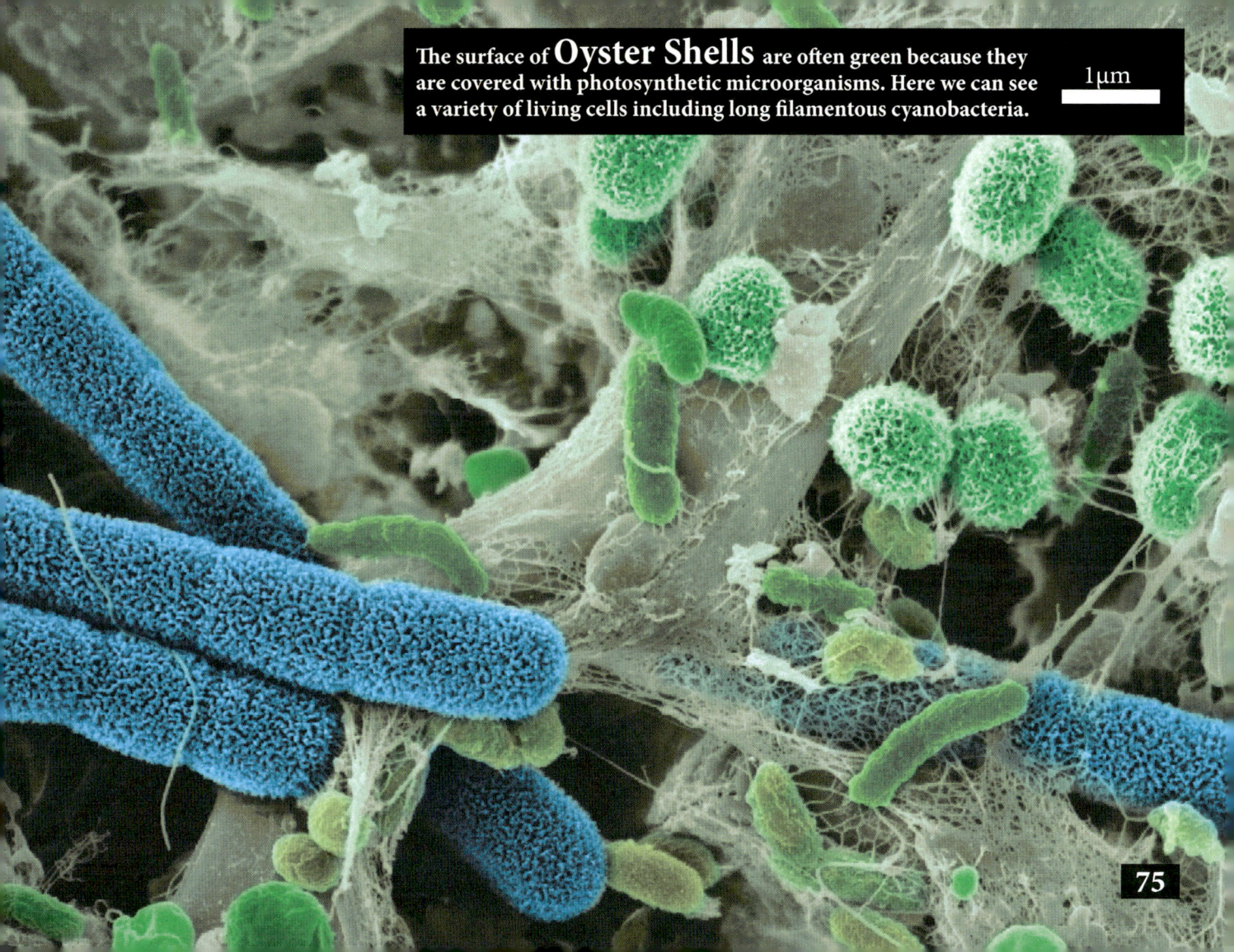

The surface of **Oyster Shells** are often green because they are covered with photosynthetic microorganisms. Here we can see a variety of living cells including long filamentous cyanobacteria.

1μm

75

Water Moss, more specifically known as Charophyta, is a common type of freshwater algae found across North America. The cellular structure consists large tube shaped cells covered in smaller chloroplasts.

Left 10μm   Right 2μm

76

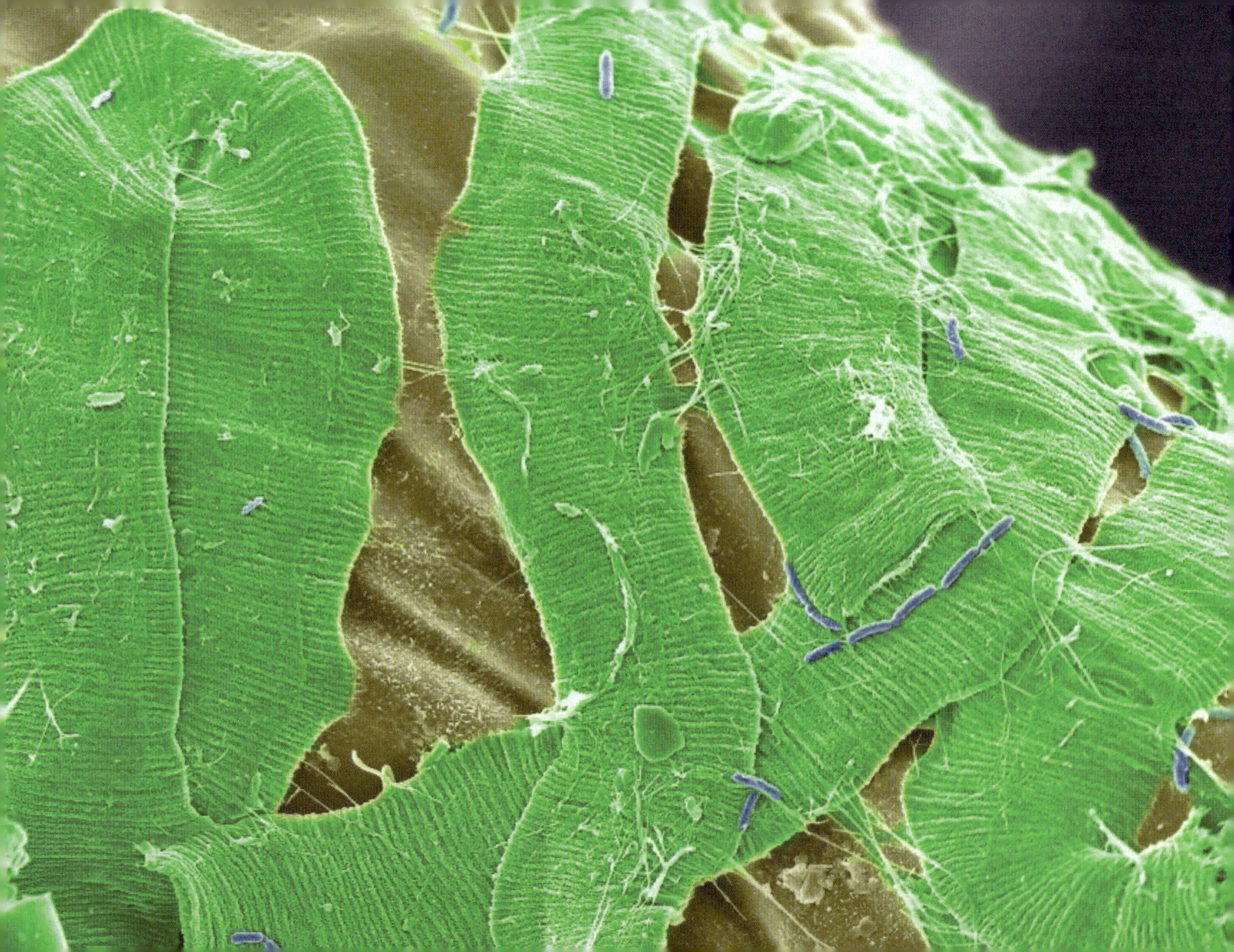

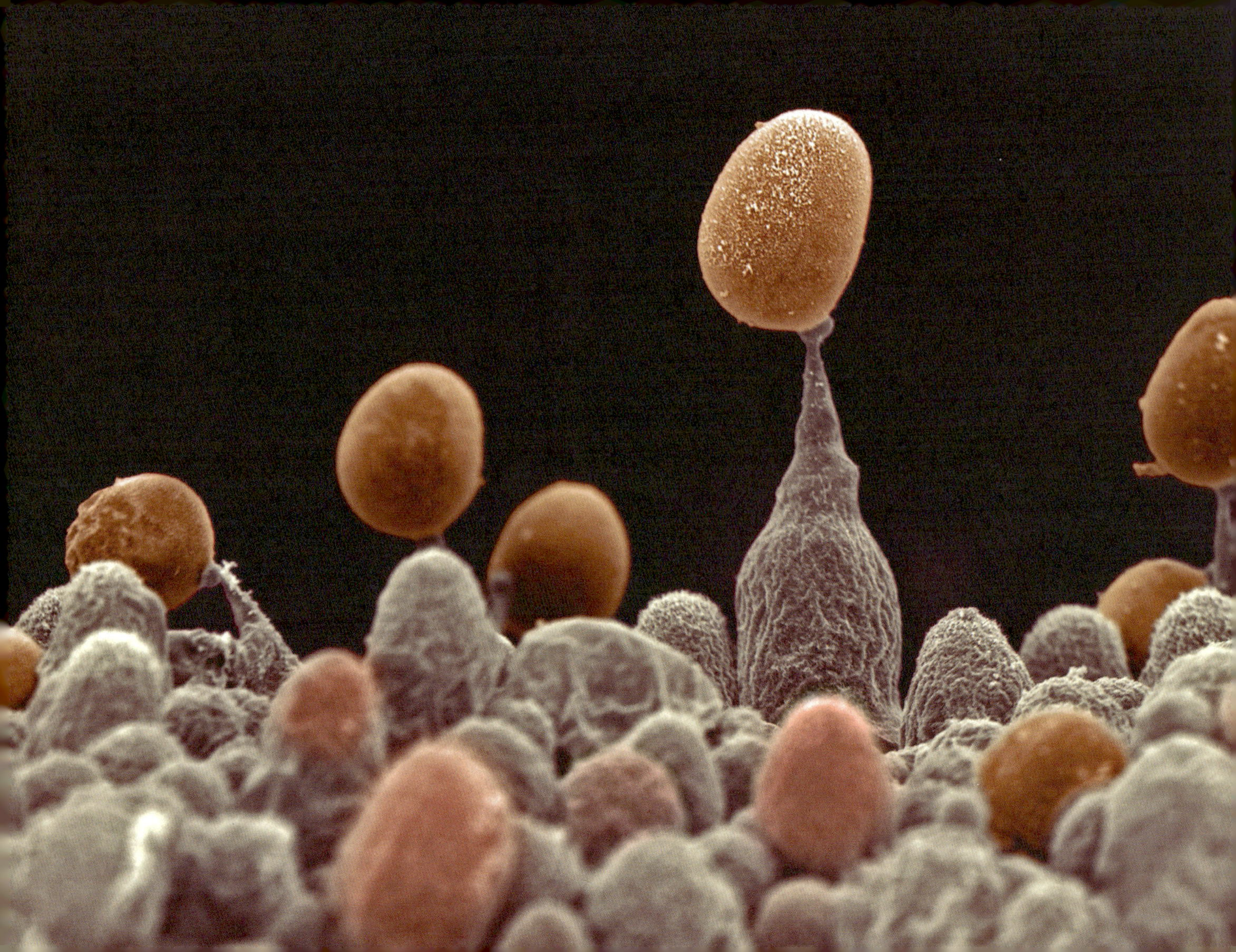

Portobello Mushroom Spores are located on the mushroom gills. Spores are the seeds of fungi, but they are so tiny that they can float away on small air currents.

3μm

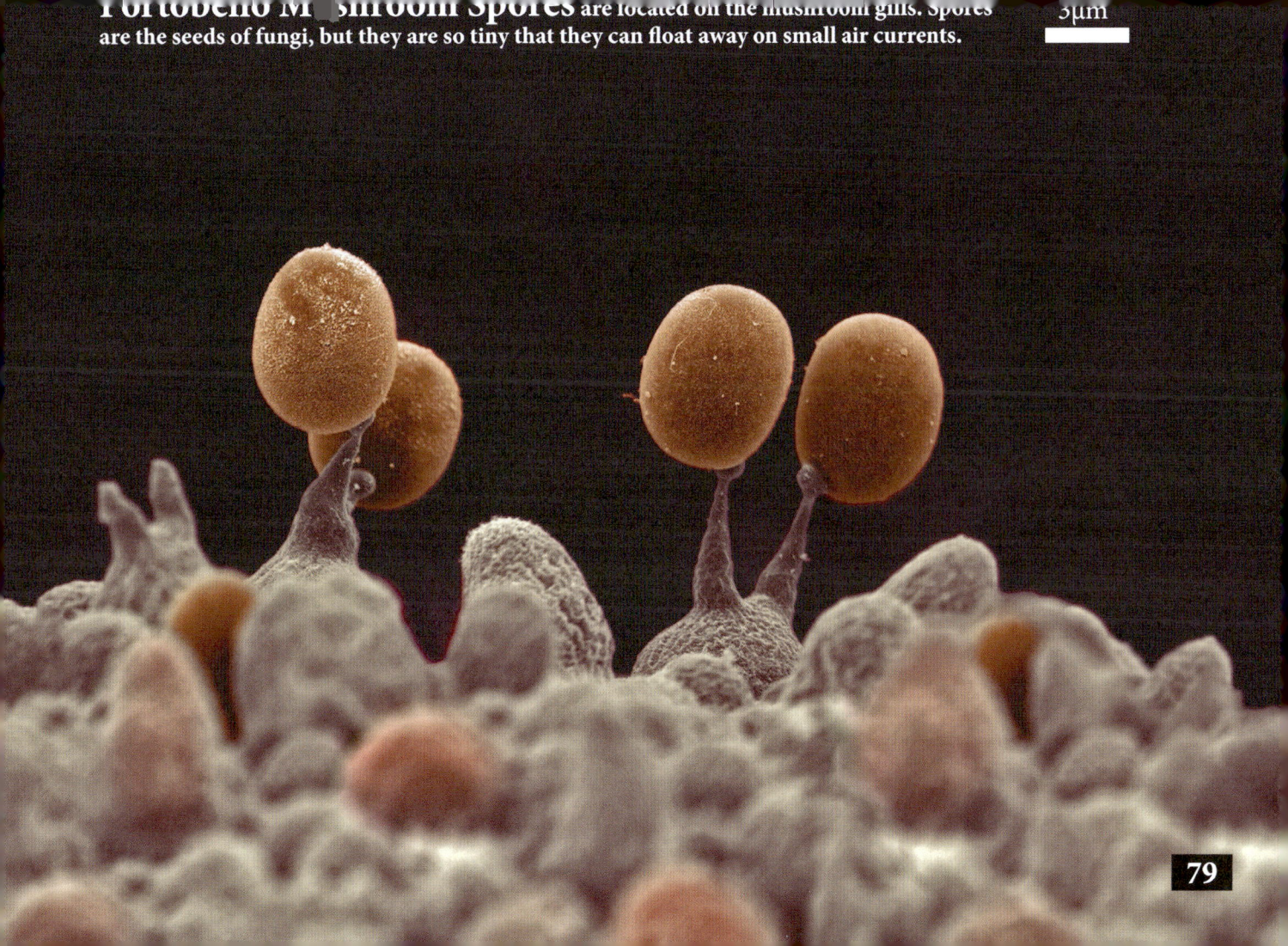

79

Cannabis plants have glandular structures called trichomes decorating the surface of the leafs and flowers. These trichomes store the psychoactive cannabinoid molecules.

Left 75μm   Right 15μm

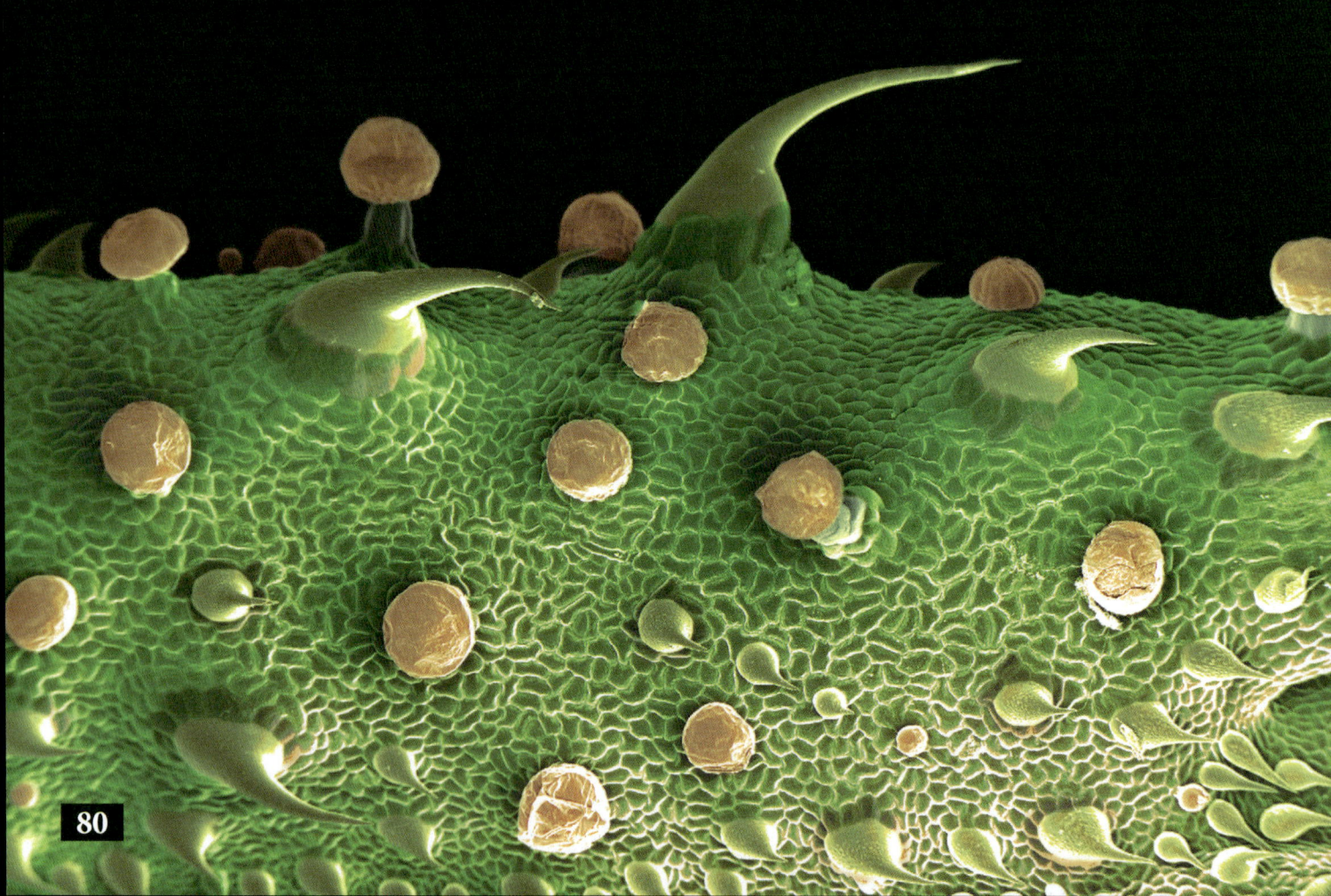

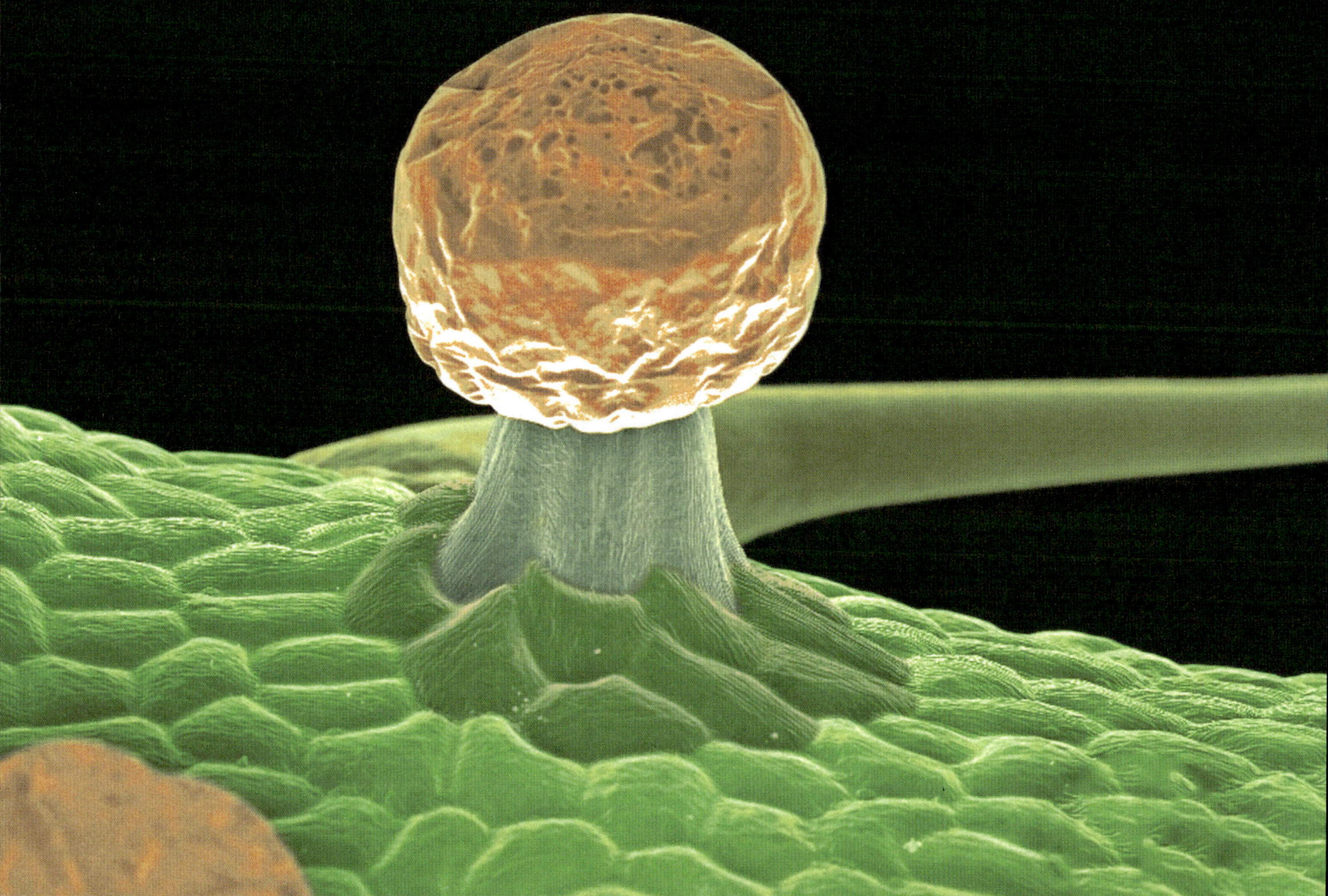

Dandelion Puffball Seed Surface
10μm

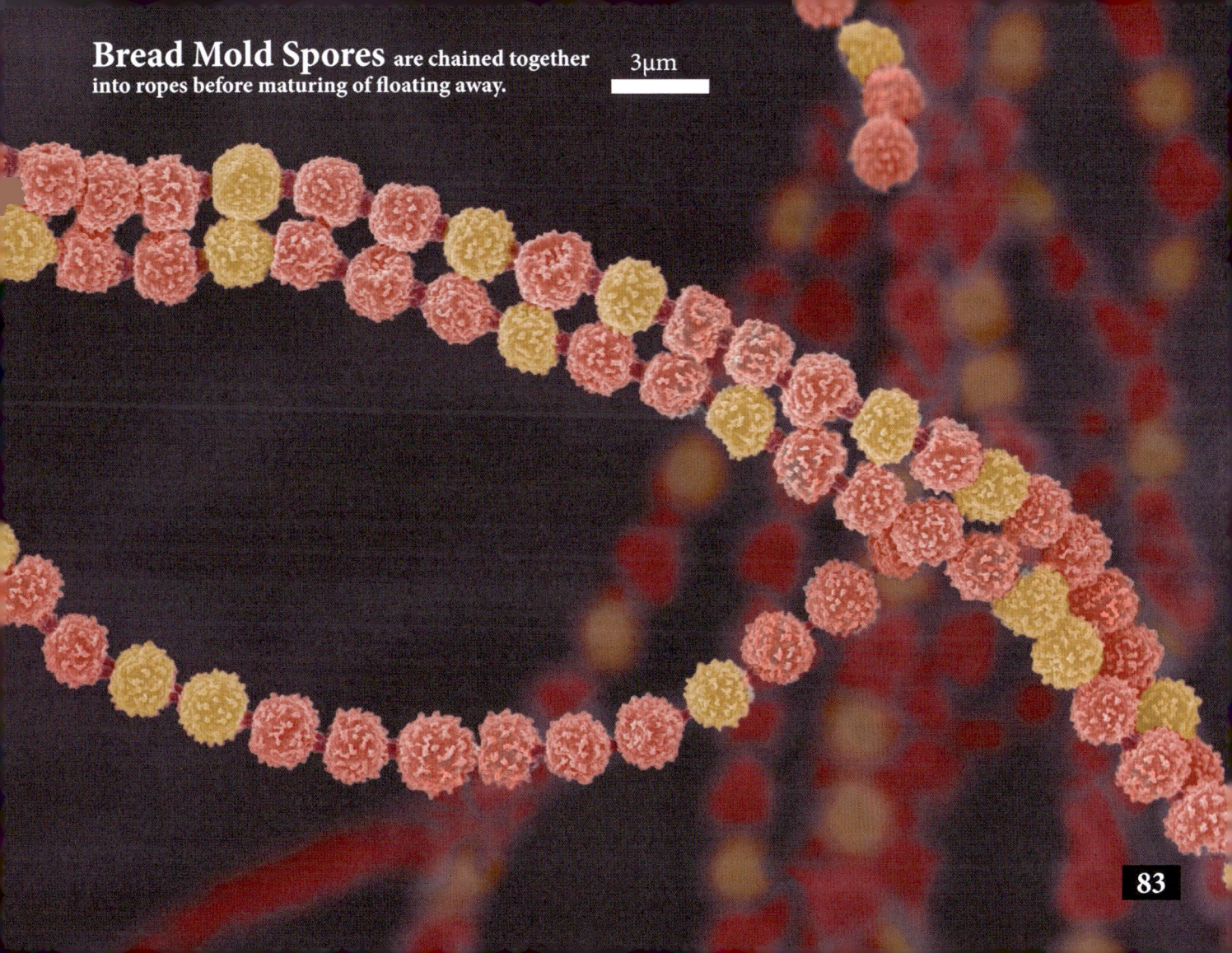

Bread Mold Spores are chained together into ropes before maturing of floating away.

3μm

MINERALS

Synthetic Diamonds

are grown using a chemical vapor deposition technique where carbon atoms are added layer by layer. Even small grain diamonds like these take days to grow under extreme conditions. Larger jewelry can be made, but the length of synthesis time makes them costly. These microscopic diamonds are used to make ultra hard cutting tools and polishing pastes.

30μm

Shown here is the broken surface of a **Black Sapphire**. Sapphire is a gemstone composed of crystalline aluminum oxide. The color is caused by mineral impurities, primarily titanium and iron.

2μm

86

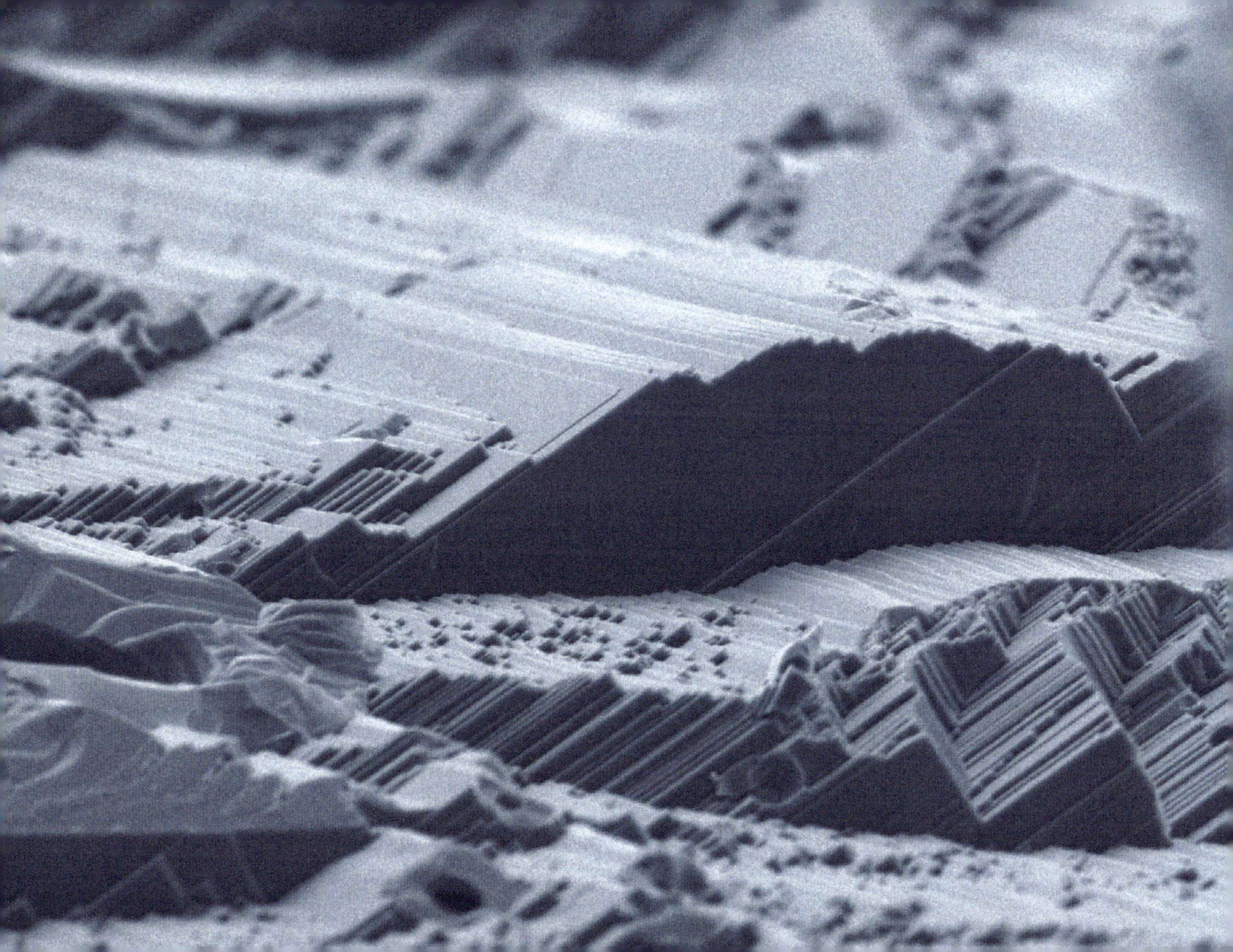

Shown here is the broken surface of a **Ruby.** Rubies differ from sapphires only by their trace impurities. Both are composed of crystalline aluminum oxide, but rubies have chromium impurities giving them their red color.

1 μm

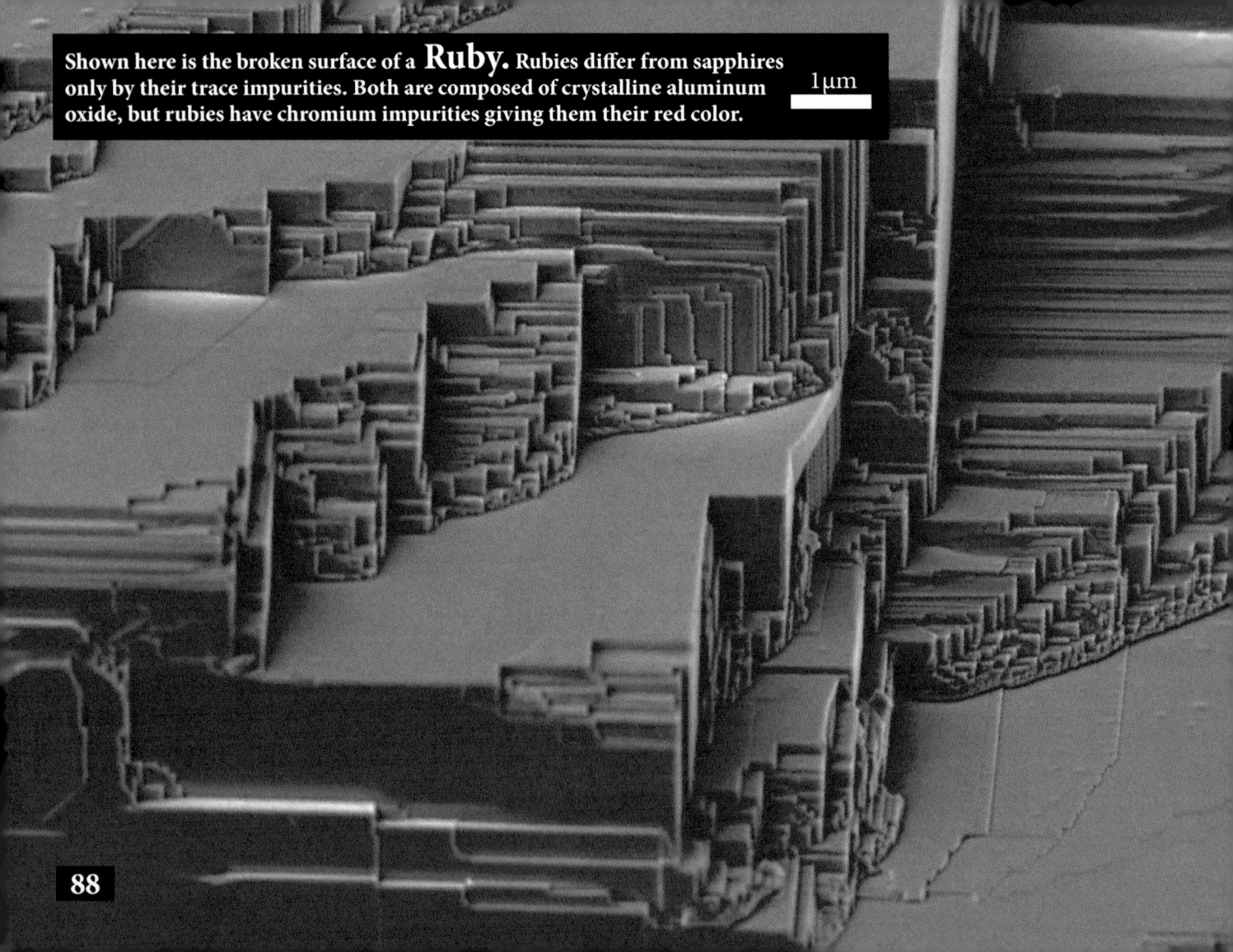

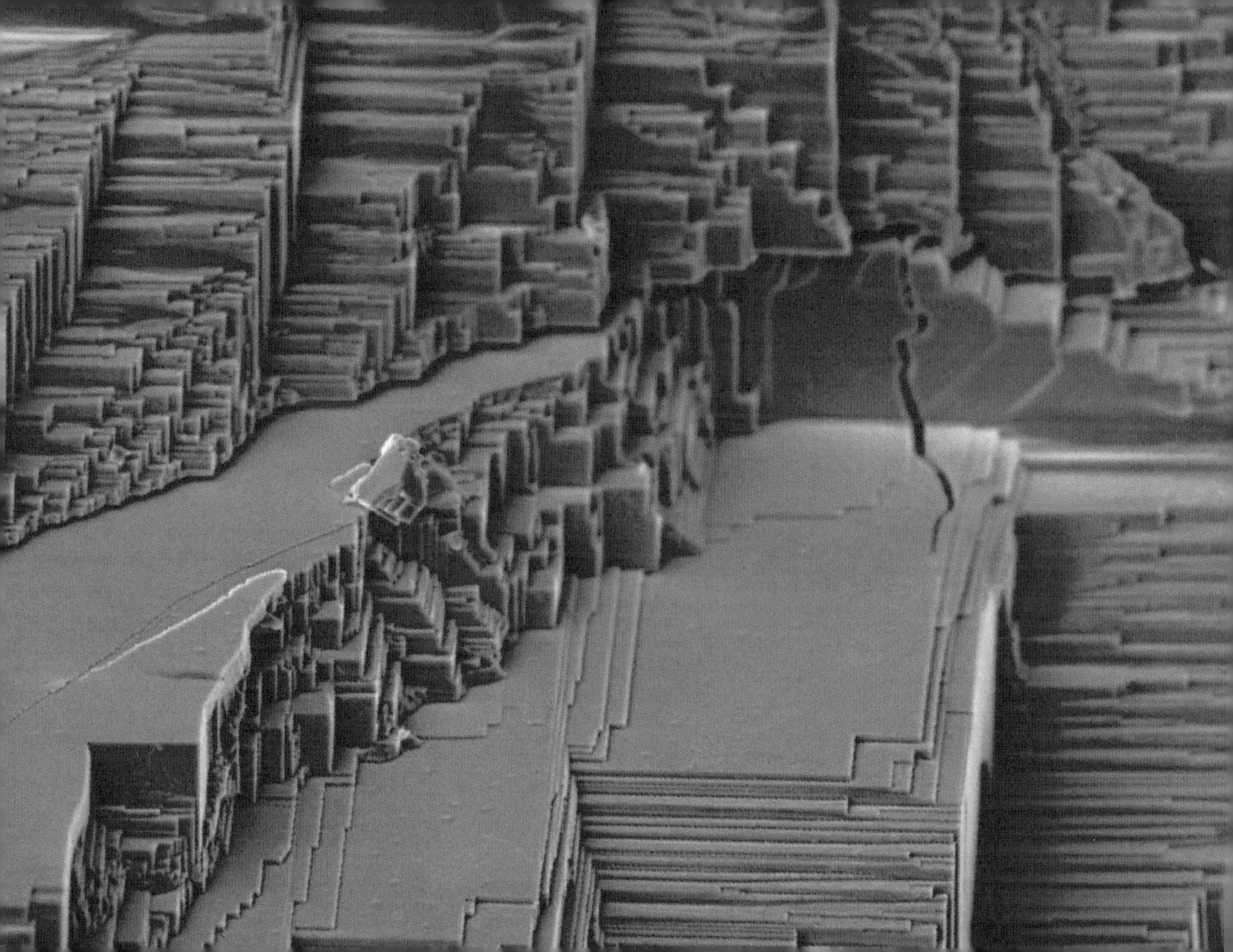

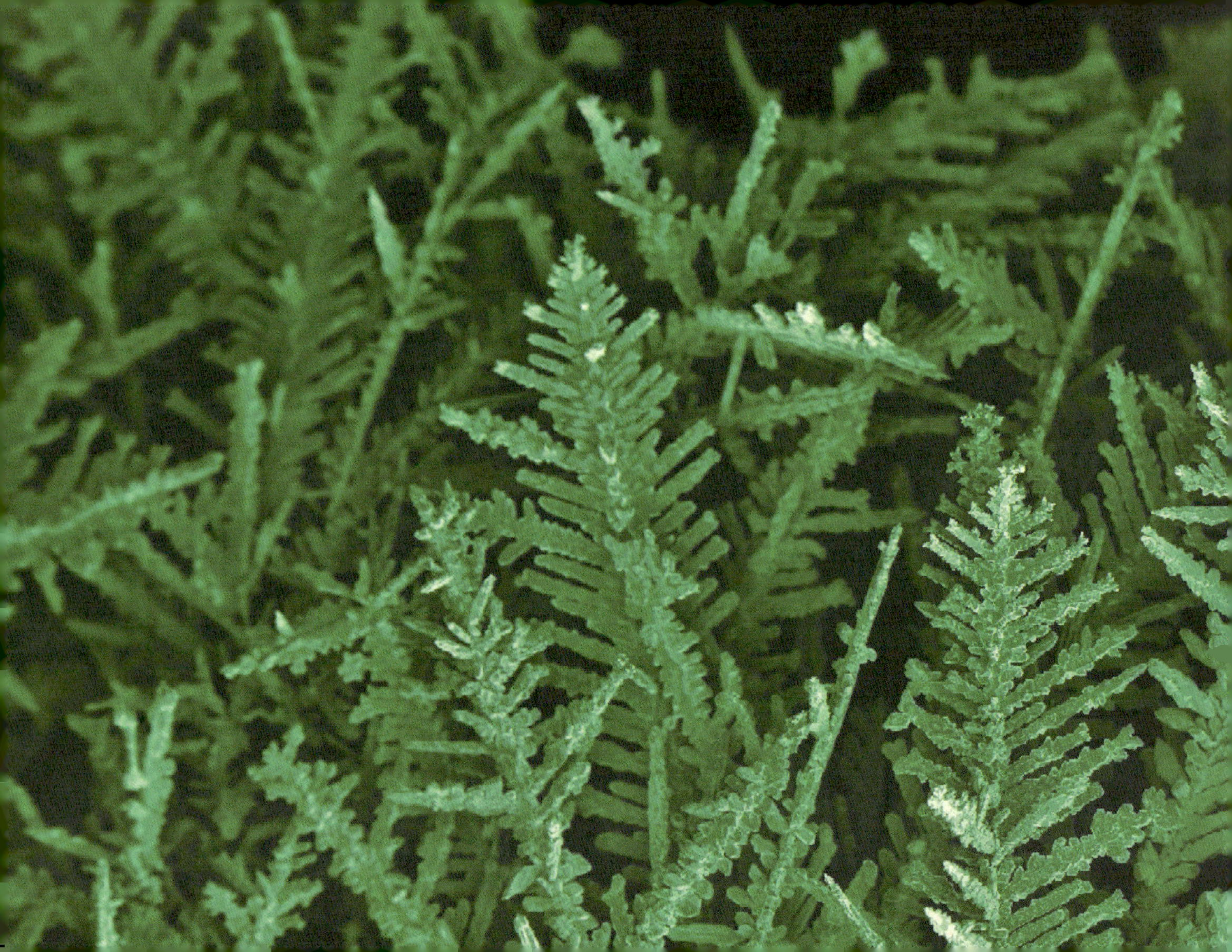

These **Silver Crystals** were grown by bringing copper metal in contact with a silver nitrate solution, a commonly studied reaction in high school.

20μm

This **Beach Rock** was particularly shiny. Viewed at this high magnification tiny smooth crystal surfaces are observed. These surfaces act like tiny mirrors reflecting the sunlight and give rocks an iridescent appearance. Left 40μm   Right 2μm

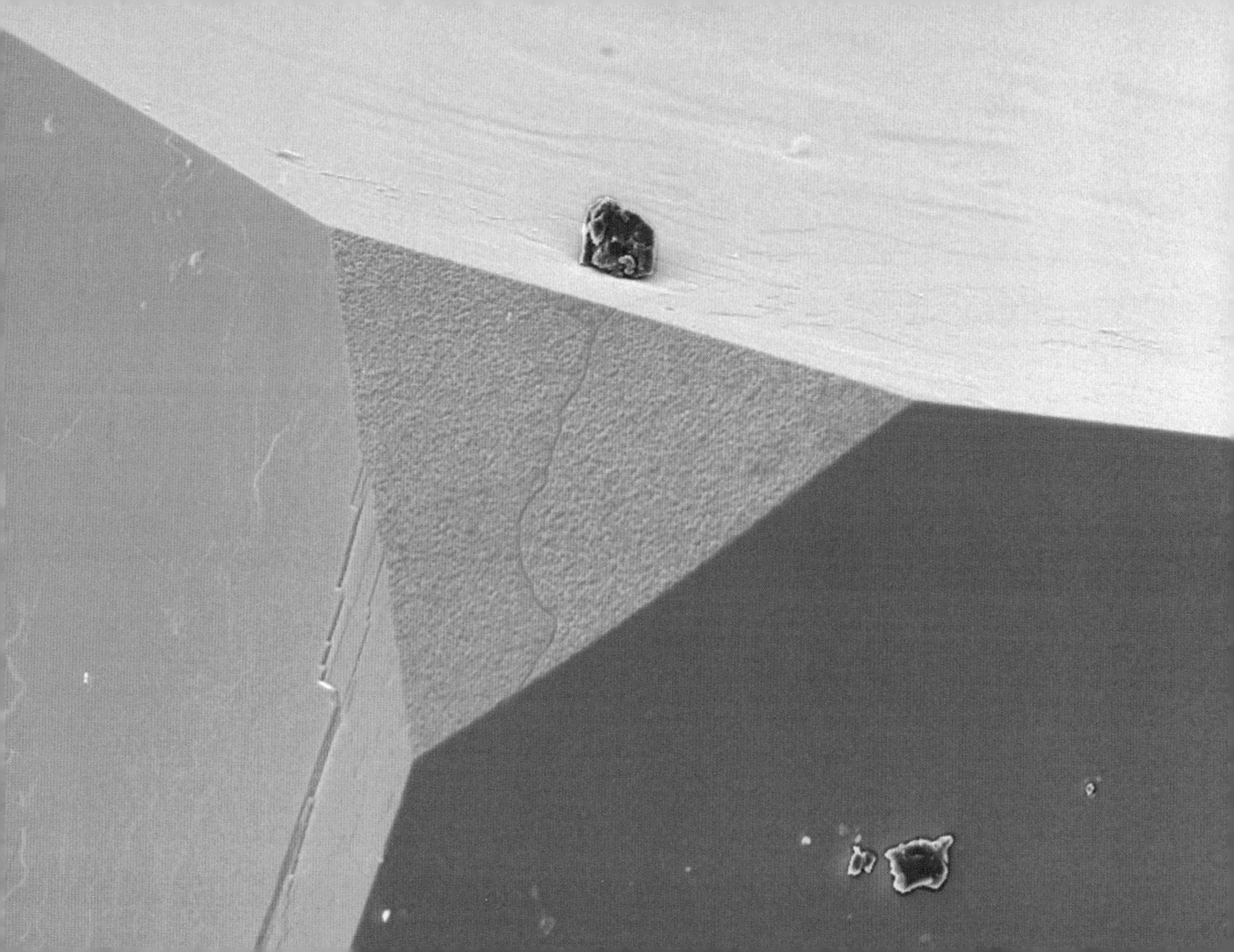

Left 20μm

Right 2μm

Potassium Phosphate
(K_2HPO_4) is a common food additive in foods like non-dairy creamer and other powdered foods.

95

Sodium Chloride (NaCl) crystallizes into a cubic lattice. However, as the salt tumbles in its container the sharp edges are eroded away like the salt grain shown here.

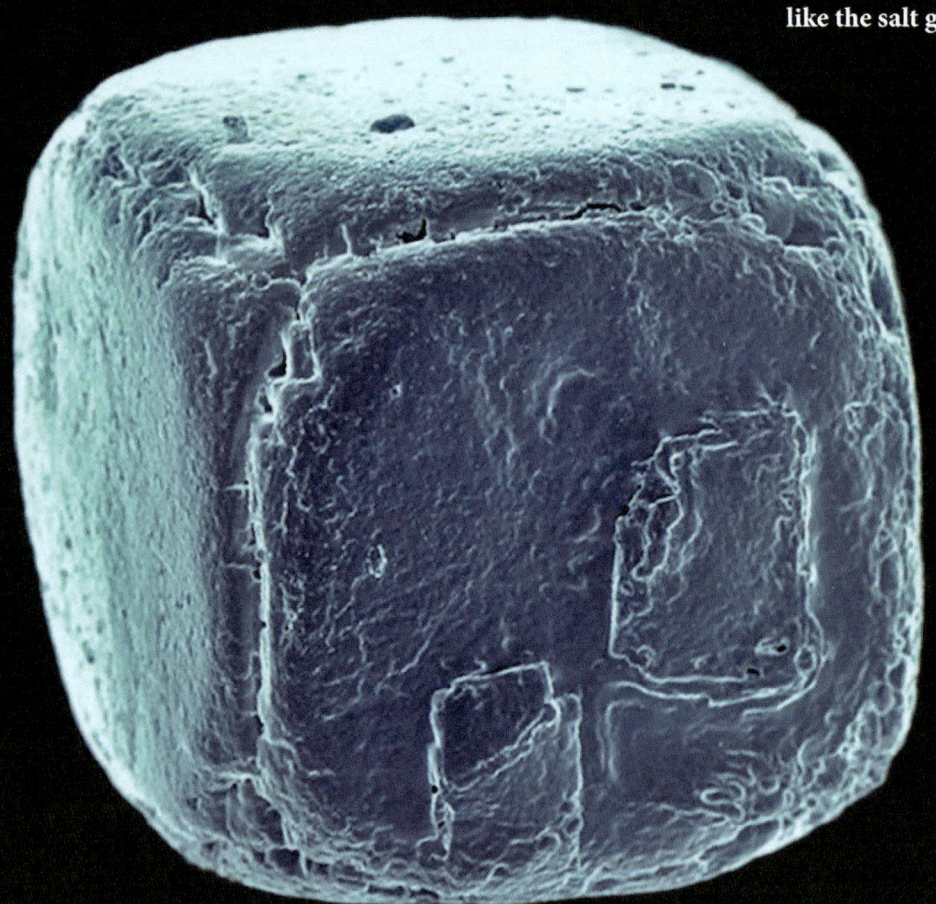

50μm

Iodized Table Salt is a version of sodium chloride with trace amounts of iodine salts introduced to prevent iodine deficiency.

100μm

Broken surface of an Amethyst Crystal. Amethyst is a form of quartz crystal with iron impurities.

5μm

Sodium Formate
(HCOONa) is commonly used as a de-icing agent at airports because it effectively prevents re-icing, even at temperatures below −15 °C

3μm

Calcium Carbonate (CaCO₃) has many ordinary applications like side walk chalk or taken as an antacid.

4μm

This is a broken surface of a **Sea Shell**. Sea shells are made of mostly Calcium Carbonate (CaCO3). Here we can see the intricate arrangement of the CaCO3 crystals. These criss-crossing layers of crystals give sea shells their strength.

1 μm

Caffeine Crystals. Many sources agree that up to 400mg of caffeine per day is safe for most adults. That is equivalent to about 4 cups of coffee.

Left 30μm

Right 2μm

103

CELLS & SCIENCE

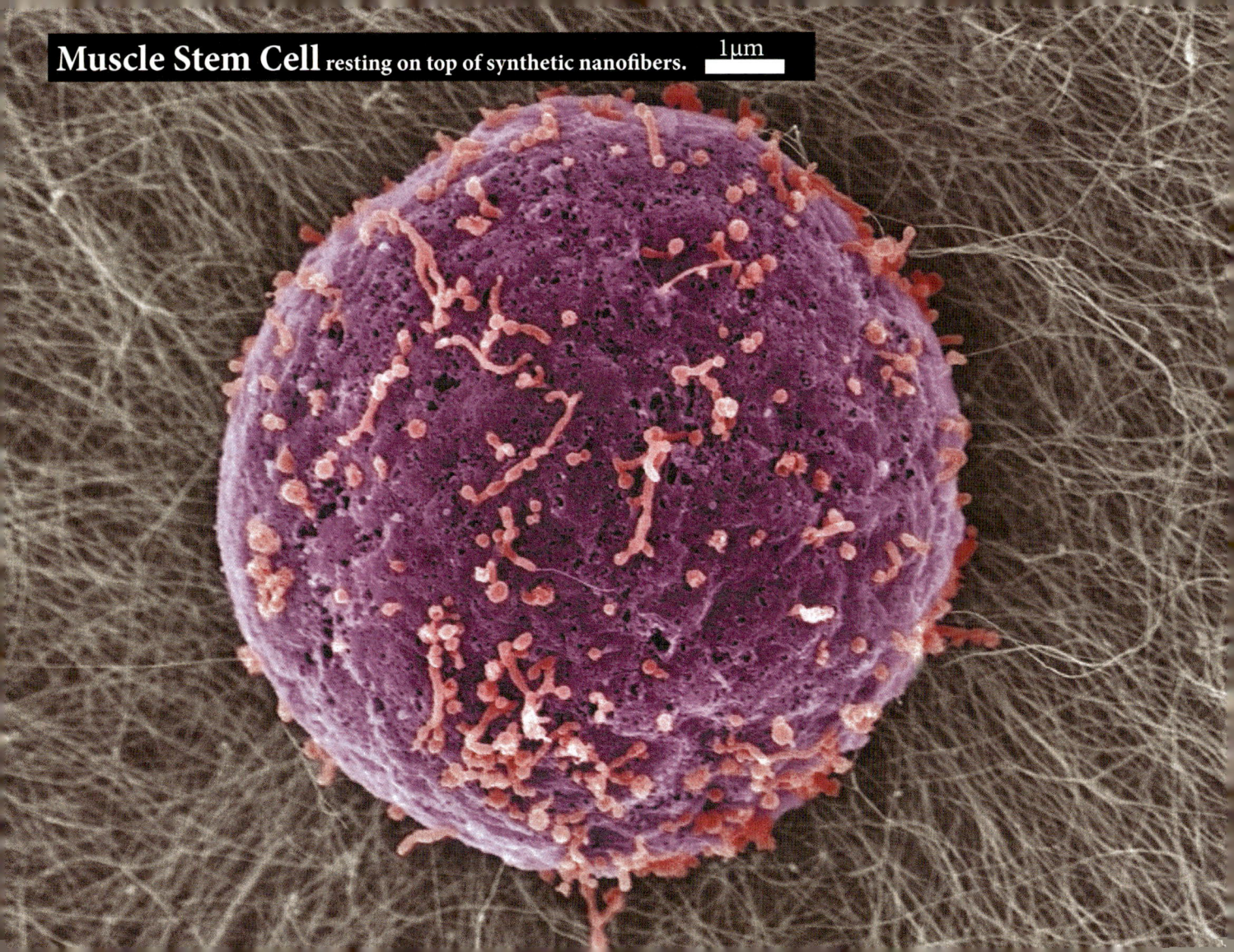

Muscle Stem Cell resting on top of synthetic nanofibers.   1μm

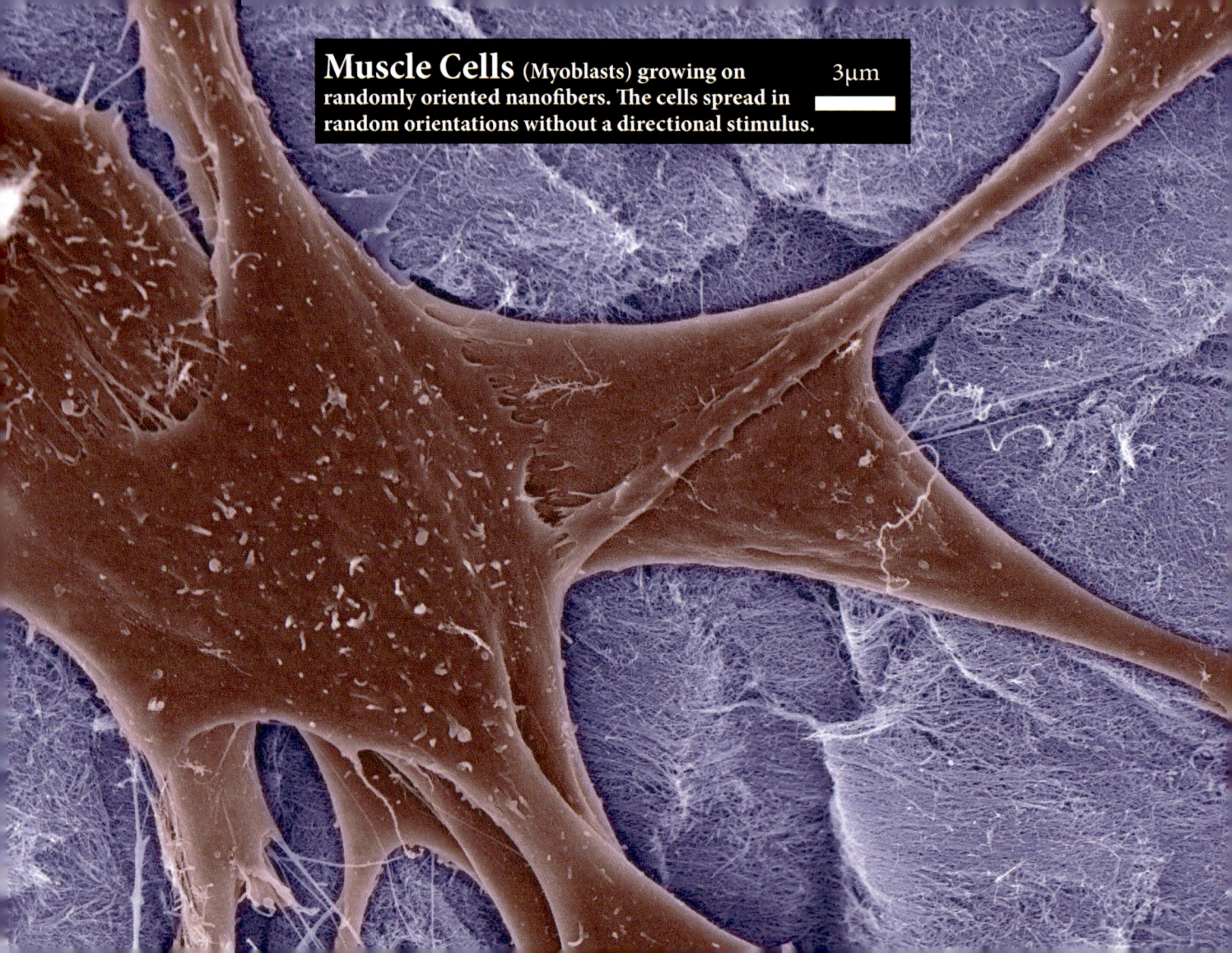

Muscle Cells (Myoblasts) growing on randomly oriented nanofibers. The cells spread in random orientations without a directional stimulus.

3μm

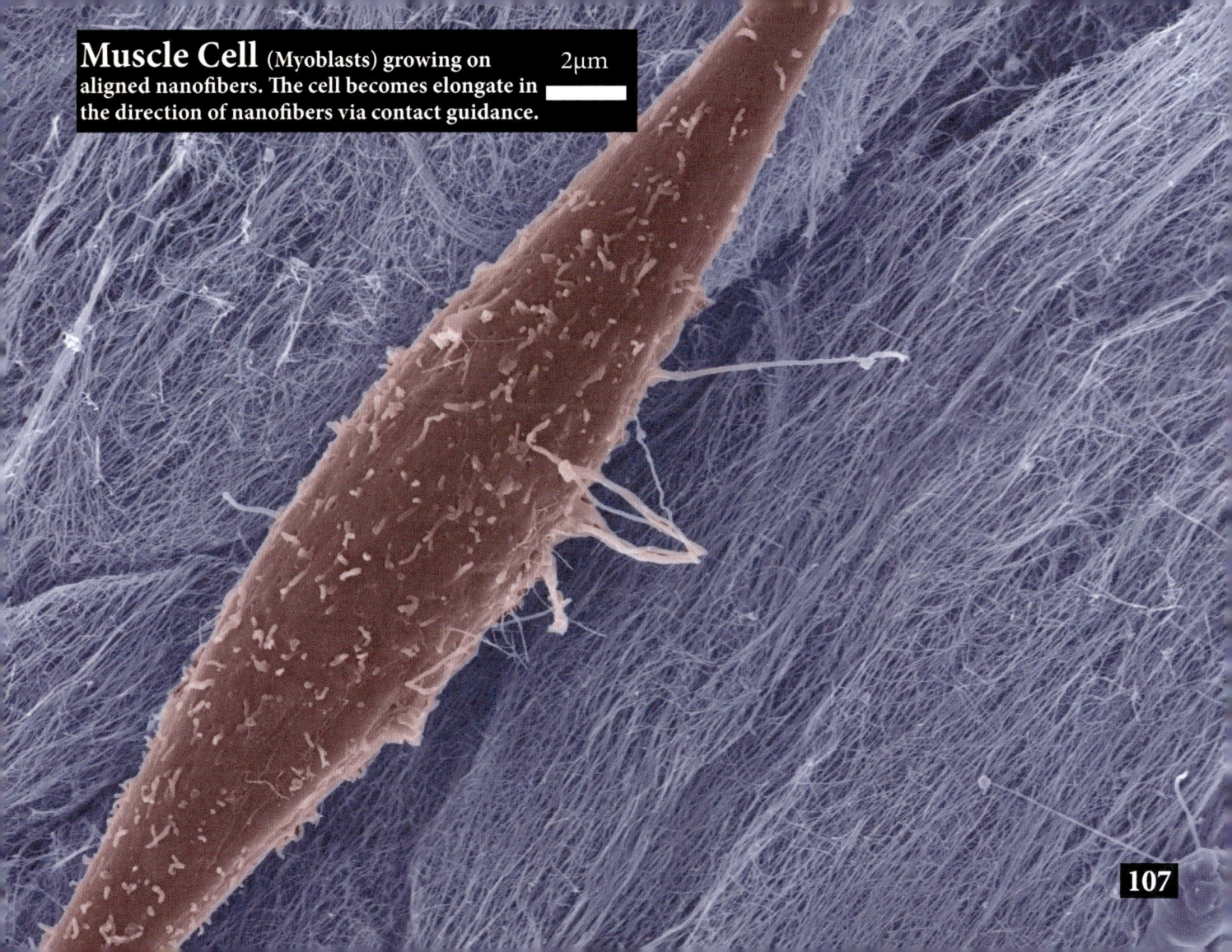

Muscle Cell (Myoblasts) growing on aligned nanofibers. The cell becomes elongate in the direction of nanofibers via contact guidance.

2μm

107

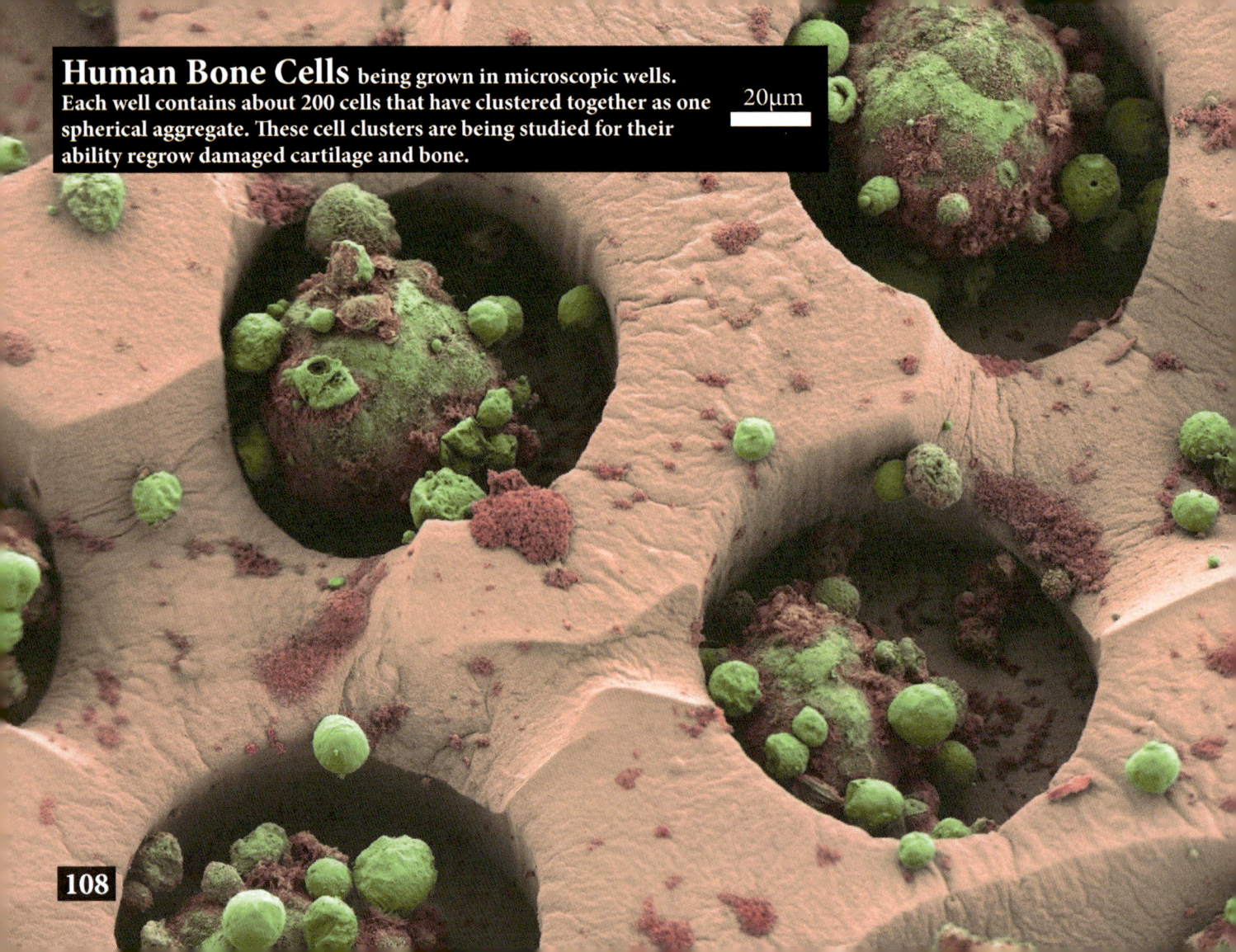

Human Bone Cells being grown in microscopic wells. Each well contains about 200 cells that have clustered together as one spherical aggregate. These cell clusters are being studied for their ability regrow damaged cartilage and bone.

20μm

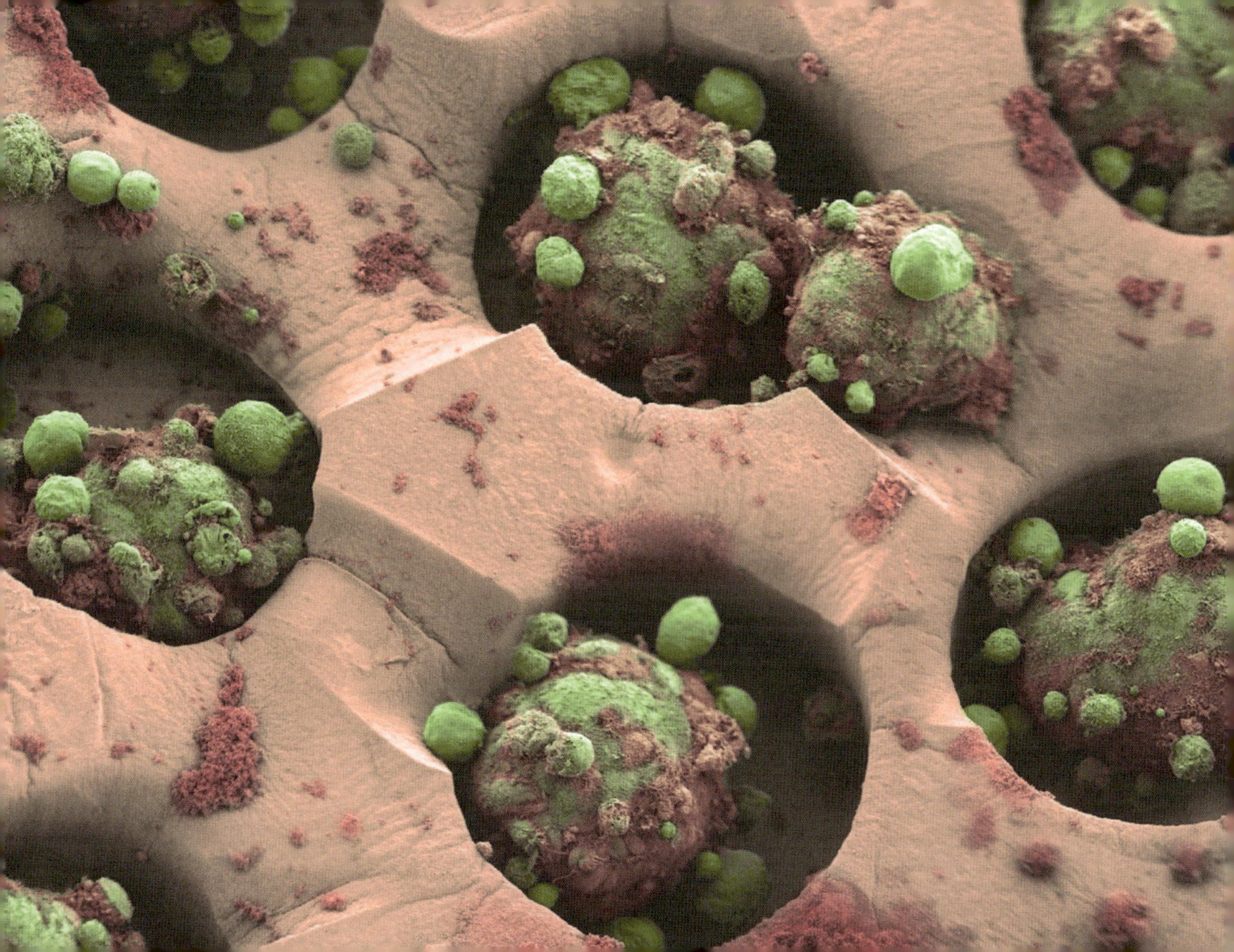

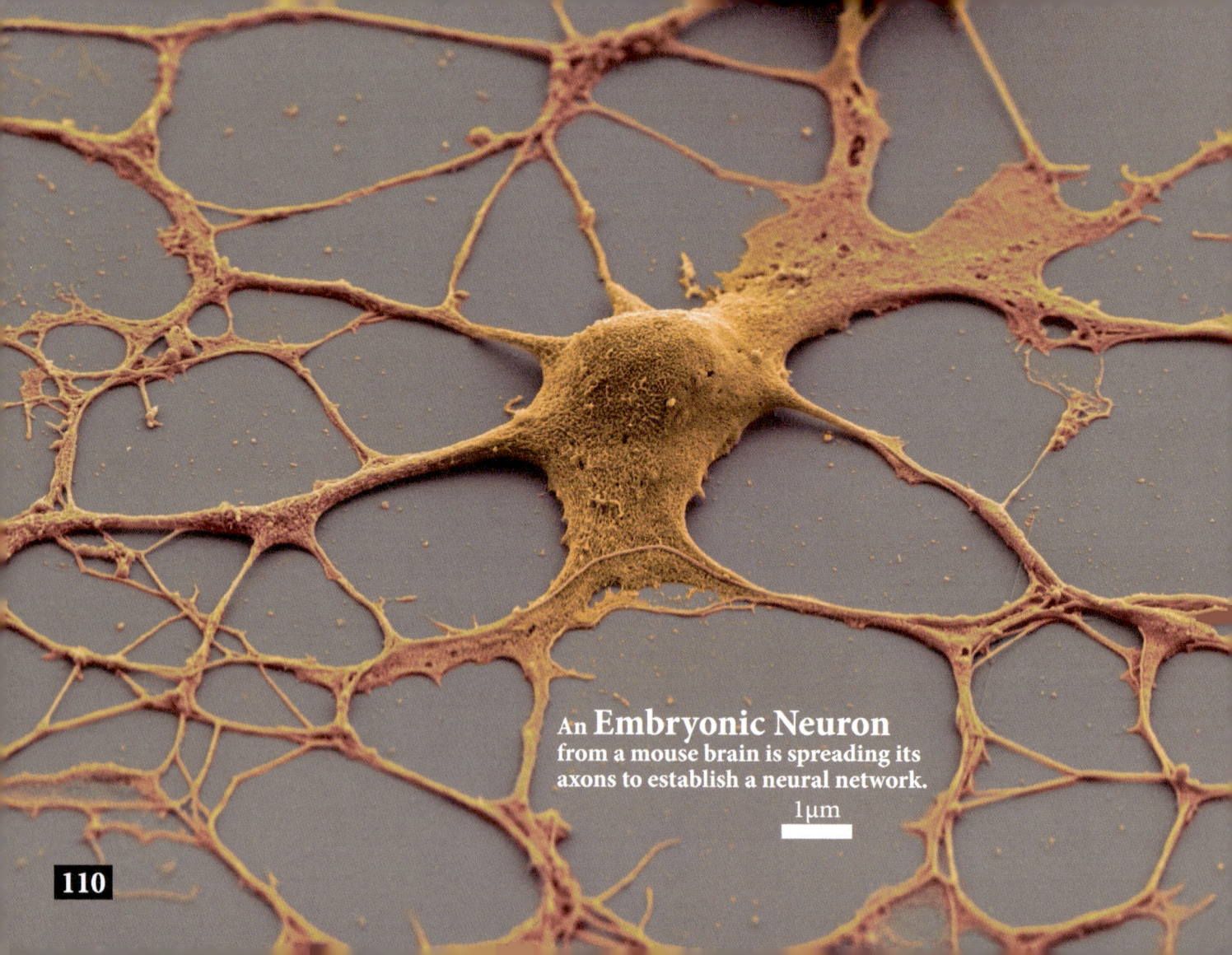

An **Embryonic Neuron** from a mouse brain is spreading its axons to establish a neural network.

1μm

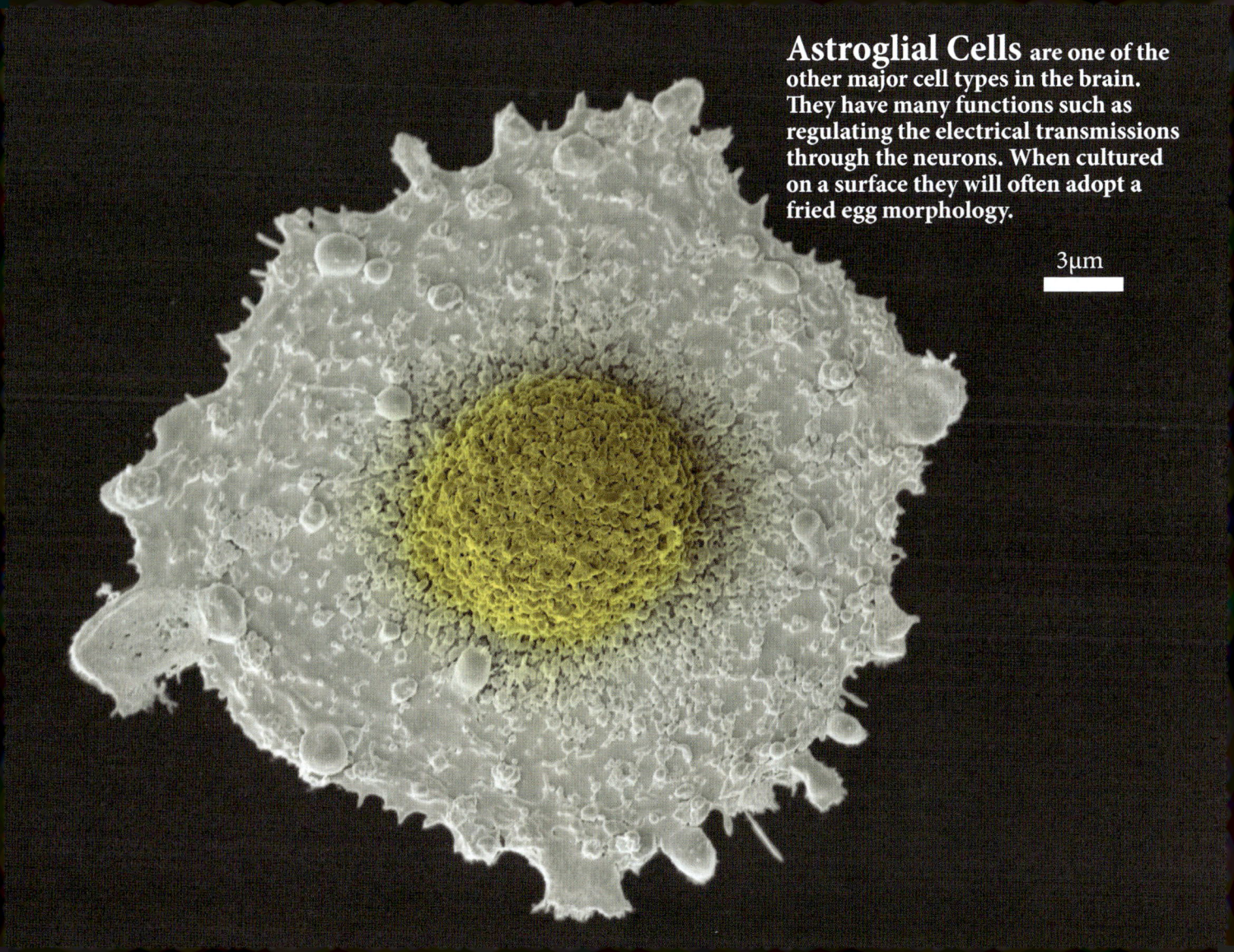

Astroglial Cells are one of the other major cell types in the brain. They have many functions such as regulating the electrical transmissions through the neurons. When cultured on a surface they will often adopt a fried egg morphology.

3μm

An **Embryonic Neuron** cultured on a synthetic nanofiber gel. These nanofibers were specifically designed to encourage nerve growth by mimicking the environment of spinal cord tissue.

1μm

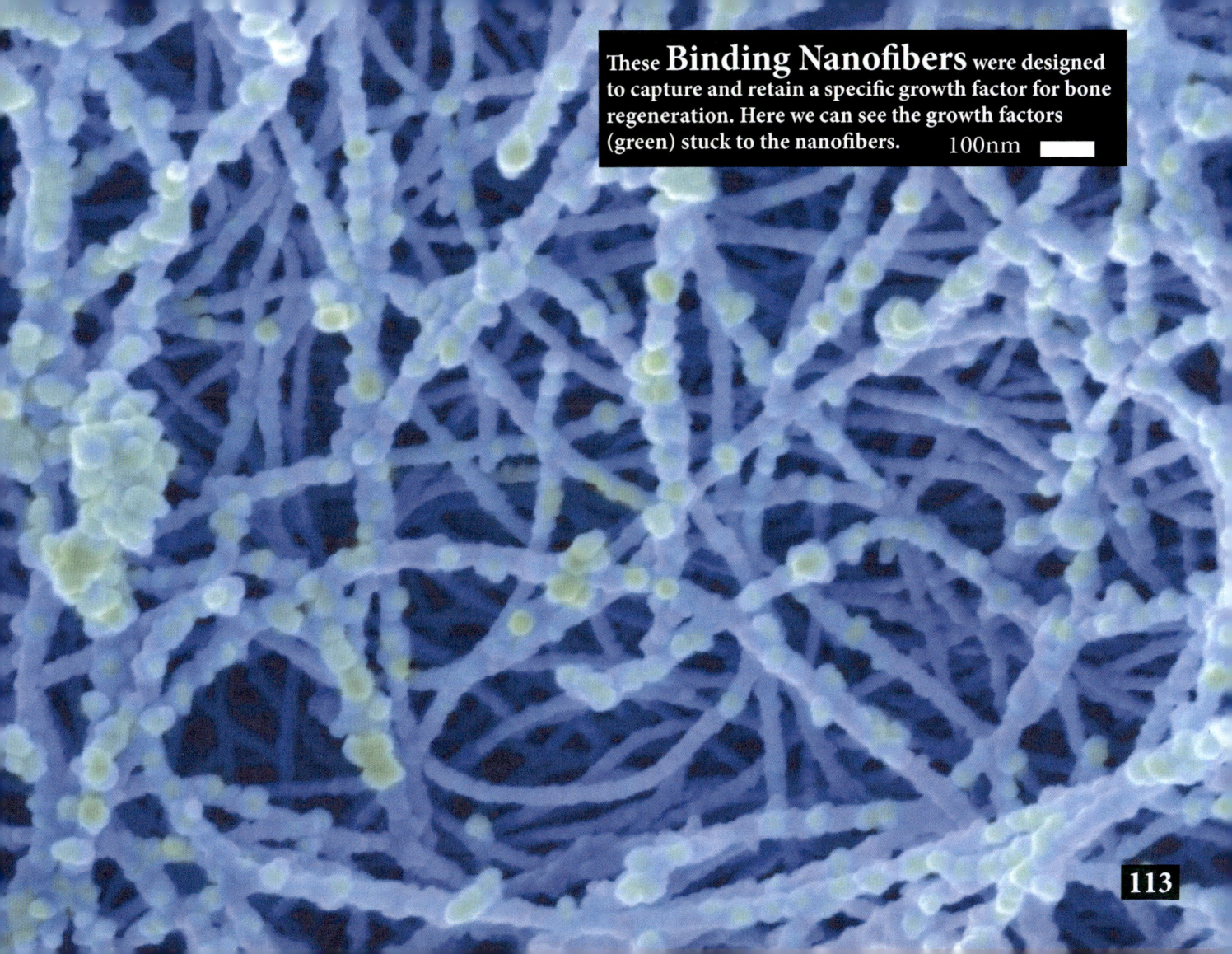

These **Binding Nanofibers** were designed to capture and retain a specific growth factor for bone regeneration. Here we can see the growth factors (green) stuck to the nanofibers. 100nm

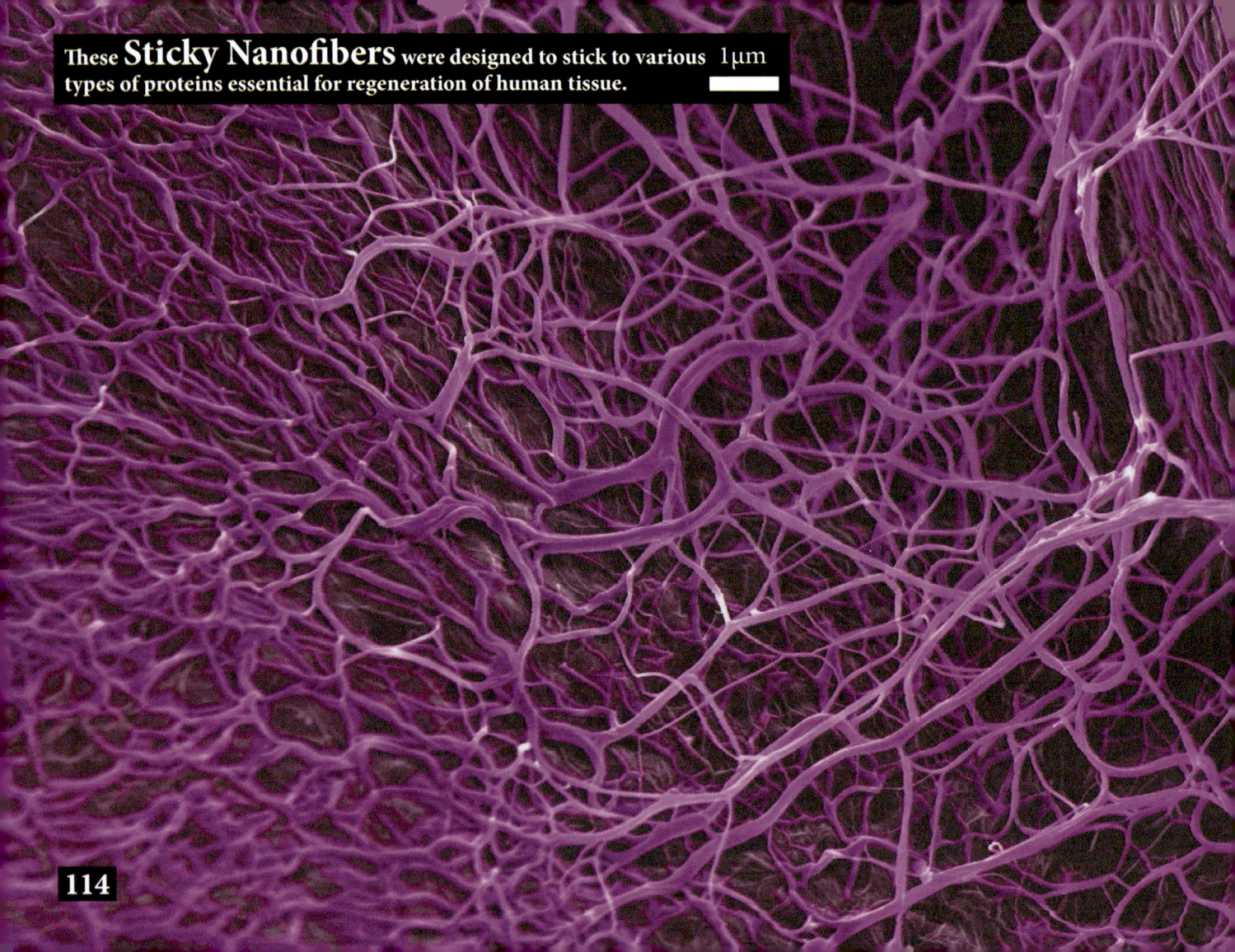

These **Sticky Nanofibers** were designed to stick to various types of proteins essential for regeneration of human tissue.

1μm

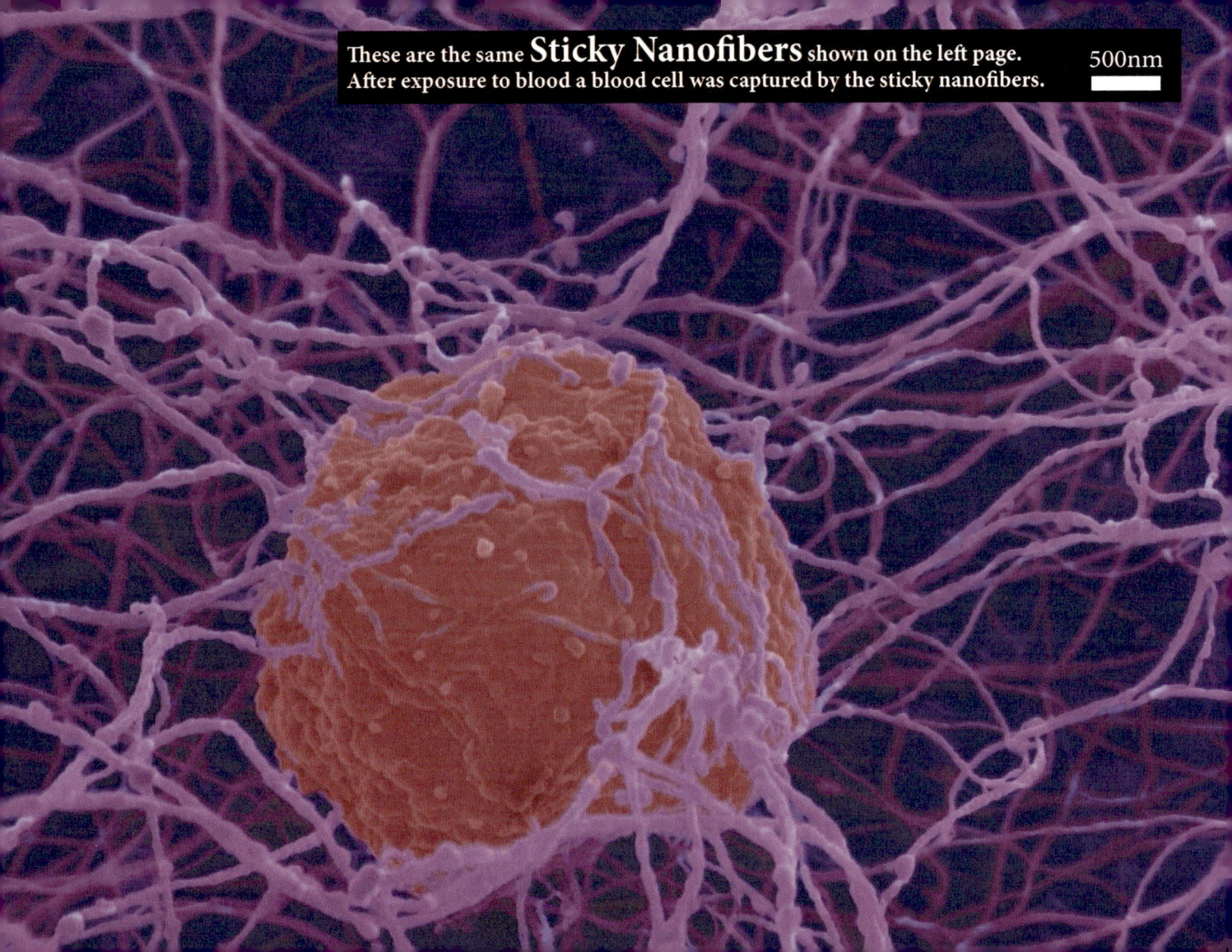

These are the same **Sticky Nanofibers** shown on the left page. After exposure to blood a blood cell was captured by the sticky nanofibers.

500nm

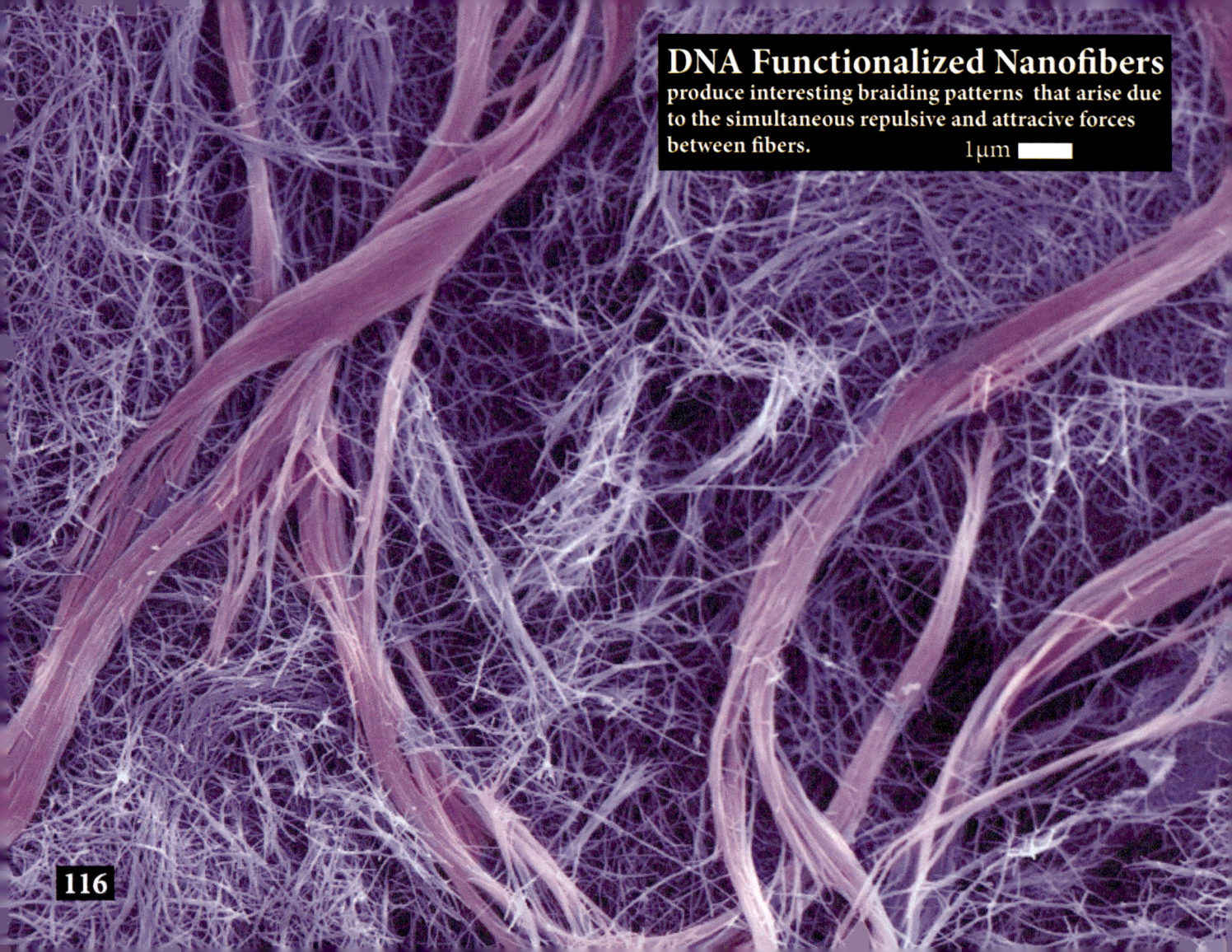

DNA Functionalized Nanofibers produce interesting braiding patterns that arise due to the simultaneous repulsive and attracive forces between fibers.

1μm

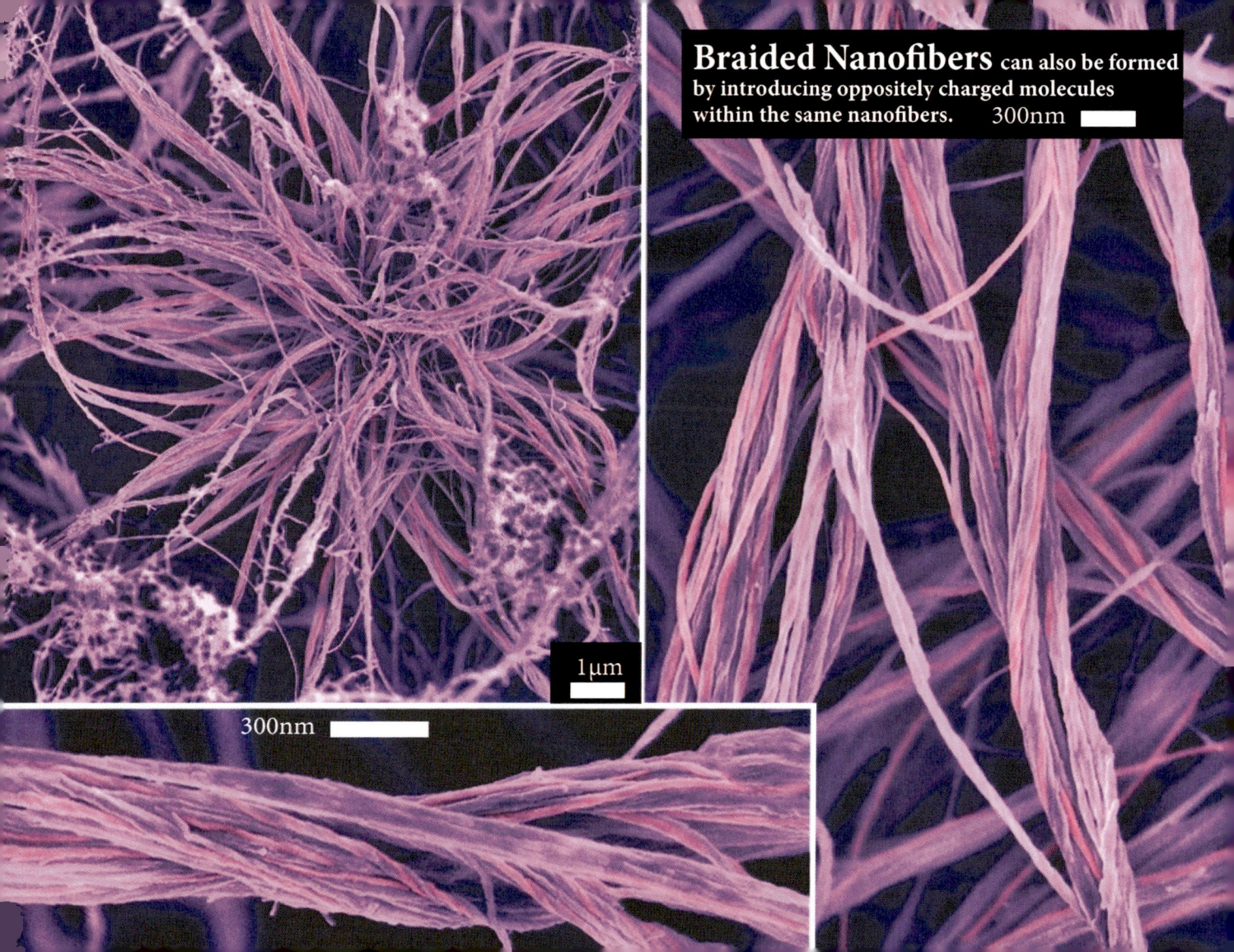

Braided Nanofibers can also be formed by introducing oppositely charged molecules within the same nanofibers. 300nm

1μm

300nm

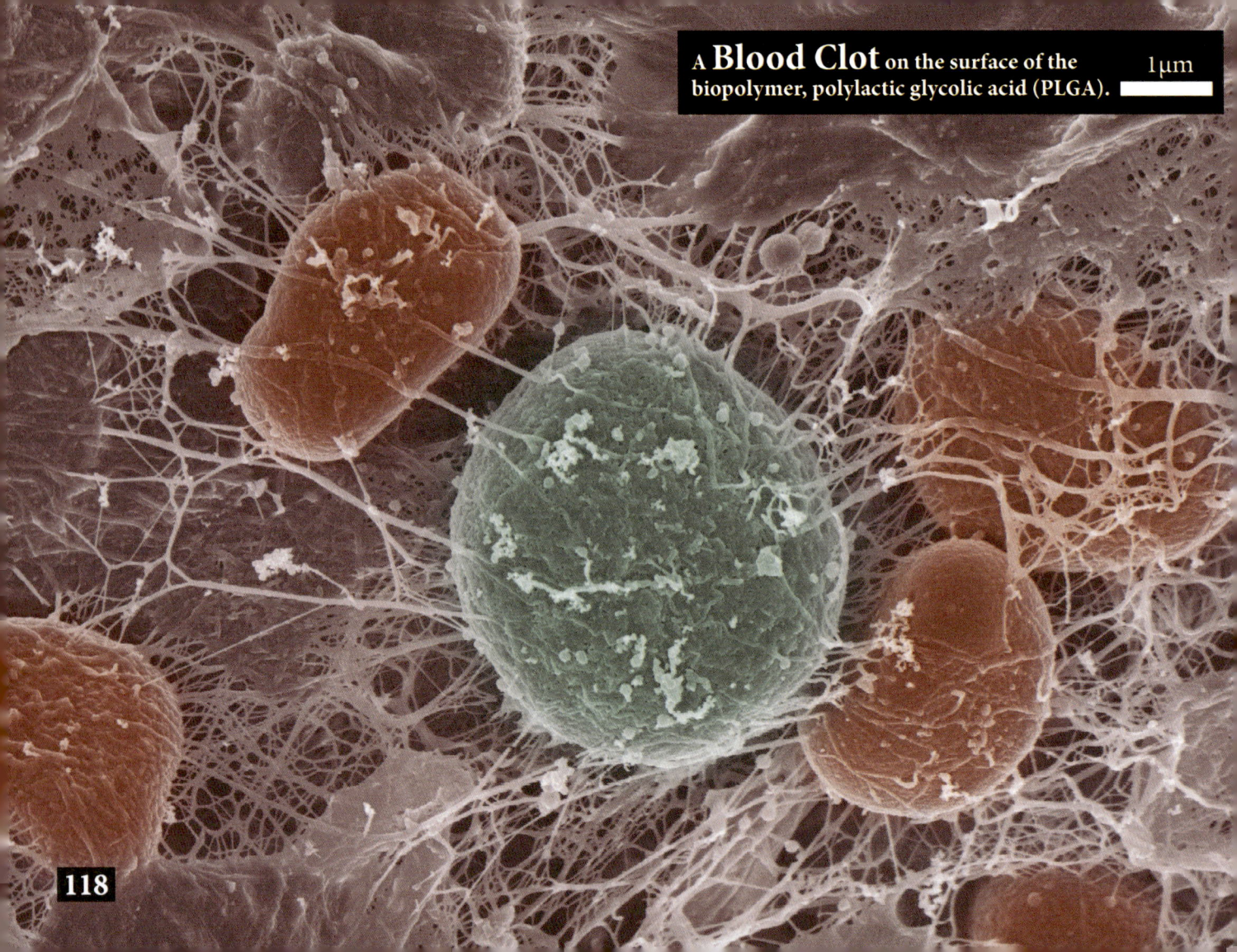

A **Blood Clot** on the surface of the biopolymer, polylactic glycolic acid (PLGA). 1μm

118

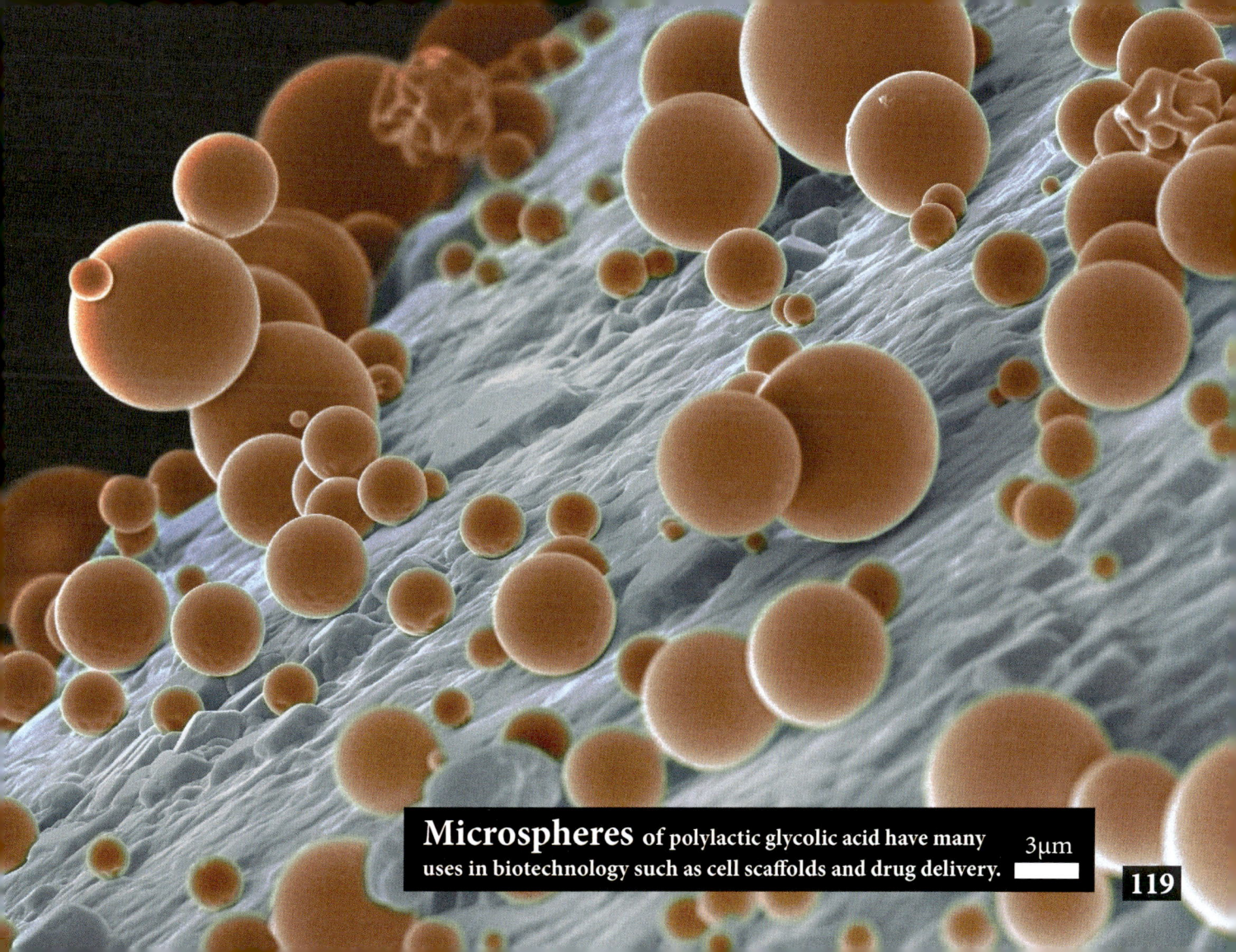

Microspheres of polylactic glycolic acid have many uses in biotechnology such as cell scaffolds and drug delivery.

3μm

About the Author

Mark Trosper McClendon

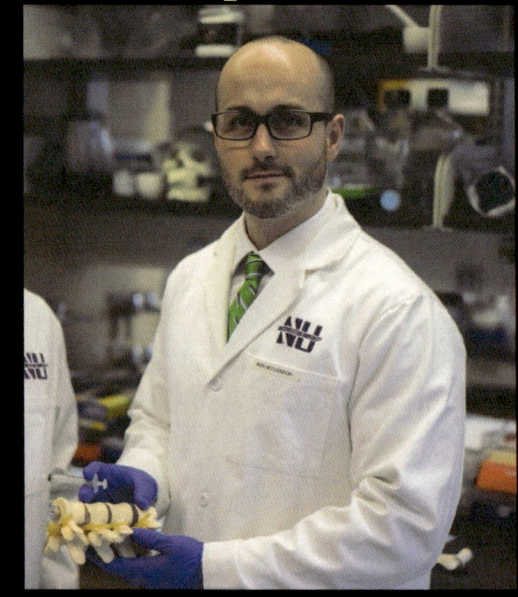

I was born and raised in Oklahoma City, OK. After earning my bachelor's degree in Chemical Engineering at the University of Oklahoma (Go Sooners!) I moved to Chicago to earn my PhD at Northwestern. I completed my graduate studies in 2014 with a focus on Nanobiotechnology, and I've continued developing nanomaterials at Northwestern's Simpson Querrey Institute. The future of medicine will be drastically influenced by nanotechnology, and translating this science from benchtop to clinical practice is my current passion. Luckily for me, this position allows me to continue taking pretty pictures of small stuff. If there is one thing I've learned from microscopy, it's that anything is amazing if you look close enough.

High Resolution Prints Available Here:
fineartamerica.com/profiles/mark-mcclendon.html

"...if we were to name the most powerful assumption of all, which leads one on and on in an attempt to understand life, it is that all things are made of atoms, and that everything that living things do can be understood in terms of the jigglings and wigglings of atoms."

Richard Feynman, The Feynman Lectures on Physics

Printed in Great Britain
by Amazon